Credits & Copyright

PYTHON MAPS
GEOSPATIAL VISUALIZATION WITH PYTHON

Adam Symington

Published by Locate Press Inc.

ISBN: 978-1-998414-07-9

Direct permission requests to `info@locatepress.com` or mail:
Locate Press, B102 5212 - 48 ST. Suite 126
Red Deer, AB, Canada, T4N 7C3

Publisher Tyler Mitchell
Editor Keith Mitchell
Interior Design Based on Tufte-LaTeXdocument class
Publisher Website `http://locatepress.com`
Book Website `http://locatepress.com/book/pymaps`

Version: 3e6ca02 (2025-02-18)

Adam Symington

Python Maps
Geospatial Visualization with Python

LOCATE PRESS

Contents

1

Introduction

1.1 Introduction

Geospatial data encompasses information linked to specific locations on Earth, serving as a cornerstone in many fields such as urban planning, environmental management, and transportation. The unique aspect of geospatial data is its diverse sources, which include satellite imagery, providing extensive coverage of Earth's surface; aerial photography, offering high-resolution images; GPS data, essential for real-time location tracking; and ground surveys, which provide detailed, localized information. These sources collectively contribute to a comprehensive understanding of the spatial dimension in various datasets, enabling a deeper insight into geographical relationships and patterns.

Visualization plays a pivotal role in geospatial data analysis. It transforms raw data into a visual context, making it easier to understand, analyze, and communicate. This is especially vital in geospatial data, where the spatial distribution of various phenomena needs to be interpreted in relation to the physical world. Effective visualization helps in identifying patterns, trends, and anomalies that might not be apparent in tabulated data, thereby enabling more informed decisions.

Python has emerged as a leading language in geospatial data analysis and visualization due to its simplicity, versatility, and powerful libraries. Libraries like Matplotlib, Geopandas, and rasterio have made it easier to manipulate and visualize geospatial data. Python's approachable syntax and supportive community make it an ideal choice for beginners and experts alike, allowing for the development of complex geospatial data analysis tools and applications.

The geospatial industry has exploded in the last decade and this has led to many companies who are generating (Earth observation companies), analysing (analytics and consultancy companies) and using (everyone) geospatial data. Humans on their own cannot hope to analyse the huge (and growing) amount of data and this means that machine learning and AI methods are increasingly being relied on. This makes Python the ideal language for anyone serious about moving into the geospatial industry due to the excellent geospatial libraries that we will go through in this book and the vast Python ML/AI ecosystem. Furthermore, the vast amount of data also necessitates advanced data engineering, something that Python is also excellent at.

This book is crafted for a diverse audience, ranging from beginners in Python to experienced data scientists and GIS professionals. It aims to provide a comprehensive guide to those interested in exploring the realm of geospatial data visualization using Python, regardless of their prior expertise in programming or data analysis. The content is structured to cater to learners at different levels, gradually building up from basic concepts to more advanced techniques.

Preview

In this book we will go through a number of examples that, together will accomplish two things, it will provide a comprehensive overview of the main geospatial libraries and concepts, and it will teach you how to produce some really interesting maps, using open source data. The fundamentals that are taught here can be used with any geospatial dataset to generate some beautiful maps or insightful analysis. In this book, we will start with four simple examples to showcase the four main datatypes that you will come across in the wild: points, lines, polygons and rasters.

We will produce a map of the world's ports and airports to showcase how to plot points and to introduce coordinate reference systems, a map of France's roads to showcase how to plot lines, a map of the Roman and Mongolian empires to showcase how to plot polygons and finally a map of distribution of World's cattle to showcase how to plot rasters. These examples will produce some interesting maps and give a strong foundation in the basics of geospatial plotting in Python, an excellent starting point for beginners.

In the next section, we will go through a number of more complicated examples that will introduce several new concepts: advanced customizations of plots, accounting for the Earth's curvature, projections, colourmaps and colour boundaries, masking and reprojecting rasters, and plotting combinations of datatypes. Through these examples we will produce maps of the World's airways, tropical storms, carbon dioxide emissions, Africa's rivers, the topography of Mars and Pakistan, and the forests of South America.

Finally, in the last section we will explore advanced concepts, how to convert a raster to points, points to a raster, polygons to a raster, a raster to polygons and a raster to lines. Throughout these examples we will produce some interesting maps showing the world's shipping lanes, bars in Europe, the topography of Greece, lighthouses of the UK and the population of Colombia.

By the end of this book you will have a solid understanding of how to use the main libraries in Python to not just plot but manipulate and analyze geospatial data in Python. We will explore coordinate reference systems, converting between data types, geospatial operations, spatial joins, projections and much more.

Library Requirements

The following libraries need to be installed in order to complete the examples in the book. This can be accomplished using pip or conda.

The library versions are also included; in the future, one or more of these libraries may be updated such that the examples in this book are not possible so it is important to lock these before starting. Furthermore, all of these examples were developed in Jupyter notebooks. These examples will work in standard Python scripts but the tutorial nature of these examples, which will require a number of exploratory plots and analysis, means that Jupyter notebooks are the ideal for these exercise.

Key Python libraries, each of which plays a distinct role in the geospatial analysis pipeline:

1. **Pandas** (*v1.4.4*): Pandas is a foundational library in Python for data manipulation and analysis, particularly suited for working with structured data. It provides data structures like DataFrame and Series, along with a wealth of functionalities to efficiently handle, clean, and analyze large datasets. In geospatial analysis,

Pandas is often used for the initial stages of data preparation and cleaning before integrating spatial components.

2. **GeoPandas** (*v0.10.2*): GeoPandas extends the capabilities of Pandas, introducing support for geospatial data. It enables handling of spatial data in the convenient form of a GeoDataFrame, integrating with Pandas' powerful data manipulation tools. GeoPandas is essential for operations like spatial joins, geospatial querying, and working with different geometric types (points, lines, polygons). It leverages libraries like Fiona for file access and Shapely for geometry operations.
3. **Shapely** (*v1.8.4*): Shapely is a library for manipulation and analysis of planar geometric objects. It is foundational in Python's geospatial stack, providing tools for creating, manipulating, and analyzing geometric shapes like lines, points, and polygons. Shapely is often used in conjunction with GeoPandas for geometric operations like buffer analysis, overlays, and calculating areas and distances.
4. **Matplotlib** (*v3.5.2*): Matplotlib is a widely used Python library for creating static, interactive, and animated visualizations. In geospatial analysis, it's often used for plotting data, creating maps, and visualizing spatial patterns. Its integration with other geospatial libraries makes it a versatile tool for creating insightful geographic data visualizations.
5. **Cartopy** (*v0.21.0*): Cartopy is a library designed for geospatial data processing in order to produce maps and other geospatial data analyses. It is built on top of Matplotlib and provides a set of tools for creating projections, performing transformations, and handling geographic datasets. It's particularly useful for creating high-level cartographic displays.
6. **NumPy** (*v1.22.4*): NumPy is a fundamental package for numerical computing in Python. It offers support for large, multi-dimensional arrays and matrices, along with a collection of mathematical functions to operate on these arrays. In geospatial analysis, NumPy arrays are often used to handle raster data or to perform batch calculations on spatial data.
7. **Rasterio** (*v1.3.2*): Rasterio reads and writes geospatial raster datasets, such as those in the commonly used GeoTIFF format. It is designed to make working with geospatial raster data more efficient and straightforward. Rasterio is essential for tasks like raster processing, analysis, and visualization, especially when dealing with satellite imagery and other raster datasets.

8. **EarthPy** (*v0.9.4*): EarthPy is a collection of IPython notebooks and Python scripts that demonstrate a variety of Earth Science related Python use cases. This library simplifies the process of downloading, analyzing, and visualizing spatiotemporal datasets. It's particularly handy for earth science students and researchers working with environmental data.
9. **RioXarray** (*v0.12.2*): RioXarray is a library built on top of Rasterio, extending its capabilities to work with xarray objects. It adds raster-specific features to xarray's multidimensional arrays, making it easier to manipulate and analyze raster data in a more Pythonic way, especially in conjunction with libraries like Dask for parallel computing.
10. **GDAL** (*v3.5.2*): GDAL (Geospatial Data Abstraction Library) is a translator library for raster and vector geospatial data formats. It is a powerful tool for data conversion and processing, supporting numerous raster and vector formats. In Python, the GDAL library is often used for complex geospatial data processing tasks and data conversions.
11. **OverPy** (*v0.6*): OverPy is a Python wrapper for the OpenStreetMap Overpass API. It allows users to query OpenStreetMap data using the Overpass QL and parse the resulting data in Python. This is particularly useful for extracting specific geographic elements and attributes from the vast OpenStreetMap database.
12. **Affine** (*v2.3.0*): The Affine library is designed to represent affine transform matrices (a key concept in linear algebra and geometry) and to perform common operations on them.

Dataset Requirements

The following datasets will be needed in order to complete these exercises. Some of these datasets are quite large so if storage is an issue I would advise downloading the resources only when you are planning on working through a particular exercise.

1. Airports and Flight Paths - Open Flight Dataset.[1] This dataset will be used to generate maps of the world's airports and flight-paths
2. Ports - World port Index[2]
3. Roads - Meijer et al, Global patterns of current and future road infrastructure, Environ. Res. Lett. 13 064006 (2018) (10.1088/1748-

[1] `https://openflights.org`

[2] `https://msi.nga.mil/Publications/WPI`

9326/aabd42)

4. Livestock - Gilbert et al. Global distribution data for cattle, buffaloes, horses, sheep, goats, pigs, chickens and ducks in 2010. Sci Data 5, 180227 (2018). (10.1038/sdata.2018.227)
5. Emissions - EDGAR v6.0 dataset.[3] The following citations are required for use of this data. Crippa, M., Guizzardi, D., Schaaf, E., Solazzo, E., Muntean, M., Monforti-Ferrario, F., Olivier, J.G.J., Vignati, E.: Fossil CO2 and GHG emissions of all world countries - 2021 Report. Crippa, M., Solazzo, E., Huang, G., Guizzardi, D., Koffi, E., Muntean, M., Schieberle, C., Friedrich, R. and Janssens-Maenhout, G.: High resolution temporal profiles in the Emissions Database for Global Atmospheric Research. Sci Data 7, 121 (2020).
doi:10.1038/s41597-020-0462-2. IEA (2019) World Energy Balances,
www.iea.org/data-and-statistics, All rights reserved, as modified by Joint Research Centre, European Commission. Jalkanen, J. P., Johansson, L., Kukkonen, J., Brink, A., Kalli, J., & Stipa, T. (2012). Extension of an assessment model of ship traffic exhaust emissions for particulate matter and carbon monoxide. Atmospheric Chemistry and Physics, 12(5), 2641–2659. doi:10.5194/acp-12-2641-2012 Johansson, L., Jalkanen, J.-P., & Kukkonen, J. (2017). Global assessment of shipping emissions in 2015 on a high spatial and temporal resolution. Atmospheric Environment, 167, 403–415. doi:10.1016/j.atmosenv.2017.08.042
6. Tropical Storms - Bloemendaal, Nadia; Haigh, I.D. (Ivan); de Moel, H. (Hans); Muis, S; Haarsma, R.J. (Reindert) et. al. (2020): STORM IBTrACS present climate synthetic tropical cyclone tracks. Version 2. 4TU.ResearchData. dataset.[4]
7. Martian Topography - Fergason, R. L, Hare, T. M., & Laura, J. (2018). HRSC and MOLA Blended Digital Elevation Model at 200m v2. Astrogeology PDS Annex, U.S. Geological Survey.[5]
8. Topography - High resolution data source - Danielson, J.J., and Gesch, D.B., 2011, Global multi-resolution terrain elevation data 2010 (GMTED2010): U.S. Geological Survey Open-File Report 2011–1073, 26 p. doi:10.5066/F7J38R2N Data is free to use.[6]
9. Forest Cover - from `https://globalmaps.github.io/ptc.html`
10. Rivers - Hydrosheds - `https://www.hydrosheds.org/products/`

[3] https://edgar.jrc.ec.europa.eu/dataset_ghg60

[4] https://doi.org/10.4121/uuid:82c1dc0d-5485-43d8-901a-ce7f26cda35d

[5] http://bit.ly/HRSC_MOLA_Blend_v0

[6] https://topotools.cr.usgs.gov/GMTED_viewer/gmted2010_fgdc_metadata.html

hydrorivers

11. Lighthouses - Open Street Map.[7] Nothing needs to be downloaded prior to starting this exercise, we will do that within the exercise.

[7] https://openstreetmap.org/copyright

12. Shipping Lanes - World Bank - https://datacatalog.worldbank.org/search/dataset/0037580
13. Population Density - Kontour.[8]

[8] https://kontur.io/portfolio/population-dataset/

14. Empires - from https://github.com/symmy596/historical-geospatial-data

2

Mapping the World's Ports and Airports

2.1 Basic Point Plotting - Airports and Ports Map

As we embark on an insightful journey through the world of geospatial analysis, understanding the role of points and multipoints becomes essential. These fundamental elements are vital in representing specific locations like ports, airports, or landmarks. In this tutorial, we will first delve into the basics and applications of points and multipoints in Python before moving on to our port and airport maps.

The key areas covered in this tutorial include:

- Forming a conceptual understanding of Points and MultiPoints
- Learning the process of generating points and multipoints using Python's geospatial libraries, and understanding how to manipulate their properties for various applications.
- Introducing techniques for effectively visualizing point data.

Shapely

Before we delve into points in Python we are going to introduce the Shapely library.[9]

[9] https://pypi.org/project/shapely/

Shapely is a Python library that provides support for the manipulation and analysis of geometric objects, specifically for the geospatial domain. The library is built on top of the GEOS C++ library, which is a powerful library for handling geometric operations. Shapely is designed to be a lightweight and easy-to-use alternative to full-fledged geospatial libraries like GDAL/OGR or PostGIS.

Shapely was first released in 2009 and has been steadily updated since then. It was created by Sean Gillies, a developer at Mapbox who was dissatisfied with the complexity of existing geospatial libraries. He wanted to create a library that was easy to use, fast, and focused solely on the manipulation of geometric objects.

Shapely provides support for all the fundamental types of geometric objects in the geospatial domain, including Points, LineStrings, and Polygons, as well as their multi- counterparts. It provides a wide range of operations on these objects, including union, intersection, and difference, as well as buffer, centroid, and convex hull calculations. Shapely also provides support for working with coordinate reference systems (CRS) and projections, which is essential for working with geospatial data. It can read and write a variety of common geospatial file formats, including GeoJSON, WKT, and Shapefiles.

Shapely fits into the Python geospatial ecosystem as a lightweight and easy-to-use library for handling geometric operations. It is often used in conjunction with other Python geospatial libraries, such as Fiona for reading and writing geospatial data and PyProj for working with CRS and projections. One of the key benefits of Shapely is its ease of use. It provides a simple and intuitive Pythonic API for working with geometric objects, making it easy for developers to quickly prototype and experiment with geospatial data.

Points

Points in Shapely are represented using the `Point` class. A Point is a zero-dimensional object that represents a specific location in space, specified by its x, y, and, optionally, z coordinates. The coordinates of a Point can be accessed using the x, y, and z properties. Points in Shapely are immutable, meaning that their coordinates cannot be changed after they are created.

Creating a `Point` in Shapely is straightforward. The `Point` class takes two or three arguments, representing the x, y, and z coordinates of the point. For example, to create a `Point` at (2, 3), we can use the following code:

```
from shapely.geometry import Point

point = Point(0, 0)
```

We can also create a Point with a third coordinate, representing its elevation.

Points in Shapely can be used for a variety of purposes. They can represent specific locations on a map, such as the location of a city or a point of interest. They can also be used as building blocks for constructing more complex geometries, for example, the path between two Points can be represented by a `LineString`, or the boundary around a set of Points could be represented by a `LineString` or a `Polygon`.

Shapely provides a number of methods for working with `Point` objects. For example, we can use the `distance()` method to compute the distance between two points, or the `buffer()` method to create a buffer around a point. Points can also be combined with other geometries using various operations, such as `intersection()`, `union()`, or `difference()`.

Overall, Points in Shapely are a fundamental building block for working with geospatial data in Python. They provide a simple and intuitive way to represent specific locations in space, and can be combined with other geometries to construct more complex shapes.

Points in Shapely have several properties that can be accessed in Python. Here are some of the most commonly used properties of Point objects:

1. `x`: The x-coordinate of the Point.
2. `y`: The y-coordinate of the Point.
3. `z`: The z-coordinate of the Point (if it has one).
4. `coords`: A tuple of the x and y (and optionally z) coordinates of the Point.
5. `geom_type`: A string representing the type of the geometry (in this case, "Point").
6. `is_empty`: A boolean indicating whether the Point is empty.
7. `is_simple`: A boolean indicating whether the Point is simple (i.e. has no self-intersections).
8. `is_valid`: A boolean indicating whether the Point is valid (i.e. has a valid shape).

To access these properties in Python, we can simply call the property on the `Point` object. For example, to access the x-coordinate of a Point object named "point", we can use the following code:

```
print("Area", point.area)
print("Length", point.length)
print("Bounds", point.bounds)
print("Longitude", point.x)
print("Latitude", point.y)
```

```
Area 0.0
```

```
Length 0.0
Bounds (0.0, 0.0, 0.0, 0.0)
Longitude 0.0
Latitude 0.0
```

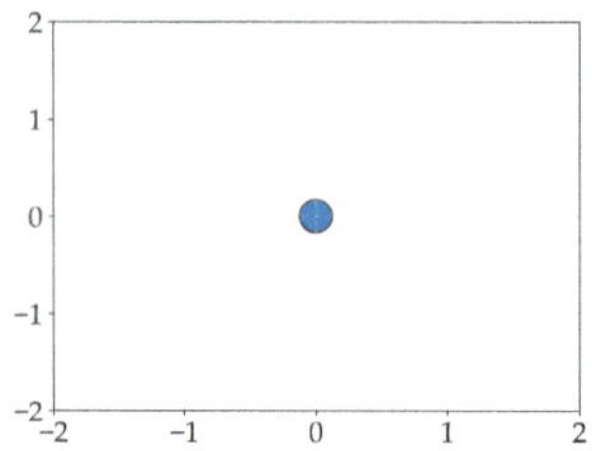

Figure 2.1: A graphical explanation of a Point geometry.

For good measure, lets plot our Point using Matplotlib, this is a visualisation book after all. The x and y coordinate can be extracted with the x and y properties of the Point.

```
import matplotlib.pyplot as plt

plt.scatter(point.x, point.y, s=500, edgecolor="black")
plt.grid(ls="--", lw=1, alpha=0.6)
plt.xticks(np.arange(-2, 3, 1))
plt.yticks(np.arange(-2, 3, 1))
plt.show()
```

MultiPoints

A `MultiPoint` is a collection of Points contained within a single object. This is useful for combining multiple Points with similar context, e.g. representing berths within the same port as a `MultiPoint`. There is also a computational aspect to the datatype in the sense that rendering say, 100 million Points, is more computationally intensive than rendering 10 MultiPoints, each containing 10 million Points because in one case we are working with 100 million separate objects and in the other we are working with 10 objects.

Constructing a `MultiPoint` is similar to constructing a `Point`. First we import the `MultiPoint` class from the shapely geometry library and then we construct a `MultiPoint` by passing a list of lists, with each sublist being a x, y (longitude, latitude pair).

```
from shapely.geometry import MultiPoint
multipoint = MultiPoint([[0, 1], [0, 2]])
```

As before we have access to a number of properties such as area, length and bonds. As we are dealing with Points, the area and length are zero but the bounds now represent the lowest/highest x and y values of the collective object.

```
print("Area", multipoint.area)
print("Length", multipoint.length)
print("Bounds", multipoint.bounds)
```

```
Area 0.0
Length 0.0
Bounds (0.0, 1.0, 0.0, 2.0)
```

Plotting a `MultiPoint` is slightly more complicated than plotting a `Point`. The *geoms* method returns a list of individual Points from the MultiPoint and then the x, y method can be used on each Point to access the coordinate. This is accomplished below with some simple list comprehension.

```
plt.scatter([geom.x for geom in multipoint.geoms],
            [geom.y for geom in multipoint.geoms],
            s=500, edgecolor="black")
plt.grid(ls="--", lw=1, alpha=0.6)
plt.xticks(np.arange(-2, 3, 1))
plt.yticks(np.arange(-1, 5, 1))
plt.show()
```

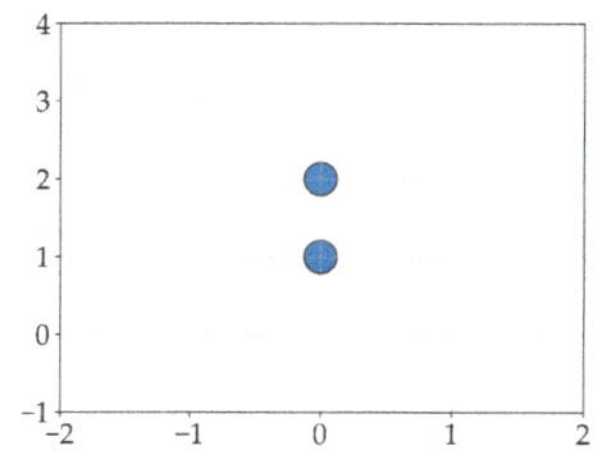

Figure 2.2: A graphical explanation of a MultiPoint geometry.

Geopandas

The final thing to go through is how to work with and plot shapely objects at scale. Shapely on its own is not suitable for large volumes of data because it works on individual geospatial objects. To process collections of geospatial data throughout this book we will use geopandas. GeoPandas is an open-source project that makes working with geospatial data in Python easier. It extends the capabilities of pandas, a popular data manipulation library, by adding support for geospatial data. The geospatial column in Geopandas is essentially a list of Shapely geometries. Consequently, all of the functionality of Shapely, which we will go through at various points in this book, is available at scale in Geopandas. GeoPandas is to geospatial data what pandas is to tabular data. It is particularly well-suited for exploratory data analysis and geospatial data processing and visualization.

Key features of GeoPandas:

1. GeoDataFrame: At the core of GeoPandas is the `GeoDataFrame`, a data structure that is similar to pandas' `DataFrame` but with additional support for geospatial data. It allows for easy handling of data that includes geographic information like coordinates, lines, and polygons.
2. Geospatial Operations: GeoPandas enables a range of spatial operations that are essential in the field of GIS. These include spatial joins, overlay, calculating areas, distances, and other geometric manipulations. It can easily handle operations like intersecting, merging, and plotting geospatial data.
3. Data Integration and File Formats: GeoPandas provides support for various file formats used in GIS such as shapefiles, GeoJSON, and KML, making it versatile in terms of data compatibility. This feature simplifies the process of reading from and writing to different geospatial data formats.
4. Visualization: The library integrates with Matplotlib to offer pow-

erful visualization capabilities. It makes plotting geographical data straightforward, with support for creating maps and adding layers of data over these maps. This is particularly useful for exploratory data analysis where visualizing data spatially is a key component.

5. Interoperability with Other Libraries: GeoPandas works well with other geospatial Python libraries such as Shapely for geometry operations, Fiona for file access, and PyProj for map projections and coordinate transformations. This interoperability makes it a central library in the Python geospatial analysis ecosystem.

This code snippet demonstrates the use of GeoPandas and Shapely in Python to create a geospatial dataframe from a set of latitude and longitude coordinates.

- `geopandas as gpd`: GeoPandas library for working with geospatial data.
- `Point from shapely.geometry`: A Shapely class to represent point objects.
- A pandas DataFrame `points_df` is created with three columns: `lat` for latitude, `long` for longitude, and `points` as an identifier for each point.
- A list geometry is created using a list comprehension. For each pair of latitude (lat) and longitude (long) from the `points_df` `DataFrame`, a `Point` object (from Shapely) is created.
- A `GeoDataFrame` `points_gdf` is created from the existing `points_df` `DataFrame`. The crs (Coordinate Reference System) is set to "EPSG:4326", which is a common geographical coordinate system (latitude and longitude). The geometry column of this `GeoDataFrame` is set to the list of `Point` objects created earlier. This column stores the geometric attributes (spatial locations) of the data.

Overall, this code is used to transform tabular data containing latitude and longitude coordinates into a `GeoDataFrame`, where each row is associated with a spatial point, allowing for spatial operations and analysis using GeoPandas and Shapely.

```
import geopandas as gpd
from shapely.geometry import Point

points_df = pd.DataFrame(
    {
        "lat": [
            1,
            2,
            3,
```

```
            4,
            5,
            6,
            7,
            8,
            9,
        ],
        "long": [1, 2, 3, 4, 5, 6, 7, 8, 9],
        "points": ["a", "b", "c", "d", "e", "f", "g", "h", "i"],
    }
)

geometry = [Point(lat, long) for lat, long in zip(points_df.lat, points_df.long)]
points_gdf = gpd.GeoDataFrame(points_df, crs="EPSG:4326", geometry=geometry)
```

```
   lat  long points                  geometry
0    1     1      a  POINT (1.00000 1.00000)
1    2     2      b  POINT (2.00000 2.00000)
2    3     3      c  POINT (3.00000 3.00000)
3    4     4      d  POINT (4.00000 4.00000)
4    5     5      e  POINT (5.00000 5.00000)
5    6     6      f  POINT (6.00000 6.00000)
6    7     7      g  POINT (7.00000 7.00000)
7    8     8      h  POINT (8.00000 8.00000)
8    9     9      i  POINT (9.00000 9.00000)
```

In the coming chapters we will go through how to create lines and polygons but for now, we are going to take the lessons learned in this opening section and apply it to some real world data to produce a map of the world's ports and airports.

Applying Points to Real-World Data: Ports and Airports

In this section we are going to produce a map showing the locations of the world's airports and seaports. This requires the use of the world port index (`https://msi.nga.mil/Publications/WPI`) and the openflights dataset (`https://openflights.org`). It is important to note that these examples assume that the files are located in the current working directory. The world port index is a dataset produced by the national geospatial intelligence agency (NGA) and contains the location, physical characteristics, facilities and services offered by major ports and terminals world-wide.

Starting with the world port index, the first step is to read the data. The data comes as a csv or as a shapefile and for the purposes of this exercise we will use the csv version because it is important to highlight how to create geospatial data from csvs. The csv contains a column for the latitude and longitude for each port and will be read into a GeoDataFrame as follows;

- Load the csv using the `read_csv` function in Pandas.

- Create a list of Shapely Points corresponding to each longitude and latitude pair.
- Create a `GeoDataFrame` by combining the original Pandas `DataFrame` and the newly created list of Shapely Points as the geometry.

```
import pandas as pd
import geopandas as gpd
from shapely.geometry import Point

ports = pd.read_csv("wpi.csv")
portgeometry = [Point(xy) for xy in zip(ports['lon'], ports['lat'])]
portgeodata = gpd.GeoDataFrame(ports, crs="EPSG:4326", geometry=portgeometry)
```

Hopefully this illustrates the point (get it) outlined in previous section that the geometry column in a `GeoDataFrame` is essentially just a list of Shapely geometries. Another thing to note is the coordinate reference system (crs) specified on the creation of the `GeoDataFrame`. The coordinates in the ports csv are in the EPSG:4326 format and hence, we specify that on the creation of the `GeoDataFrame`.

Once the data is captured within a `GeoDataFrame` it is incredibly easy to plot because Pandas/GeoPandas has plotting functions that create a matplotlib plotting object. Plotting is simple and called with the plot method on the GeoDataFrame. The `GeoDataFrame.plot()` method can be called directly to generate a plot or a matplotlib axes object can be created and passed to the plot function. This allows matplotlib methods like `set_ylim` etc to be used to customize the plot.

Geopandas plots points using a scatter plot, lines as a normal line plot and polygons using the fill method. The data can be plotted as a scatter plot to show the location of the worlds ports, or at least the ports included within this dataset. As usual with anything on a global scale this map largely just becomes a proxy for population density but it is still interesting to see the distribution of the worlds ports.

```
import matplotlib.pyplot as plt

fig, ax = plt.subplots()
port_geodata.plot(ax=ax)
plt.show()
```

This produces a generic plot because everything is set to default values (see figure 2.3, on the facing page). This is a nice way to explore the data graphically but is not particularly appealing or useful. One thing that is particularly important to change is the projection. The default projection uses what ever coordinate reference system has been set in the `GeoDataFrame`.

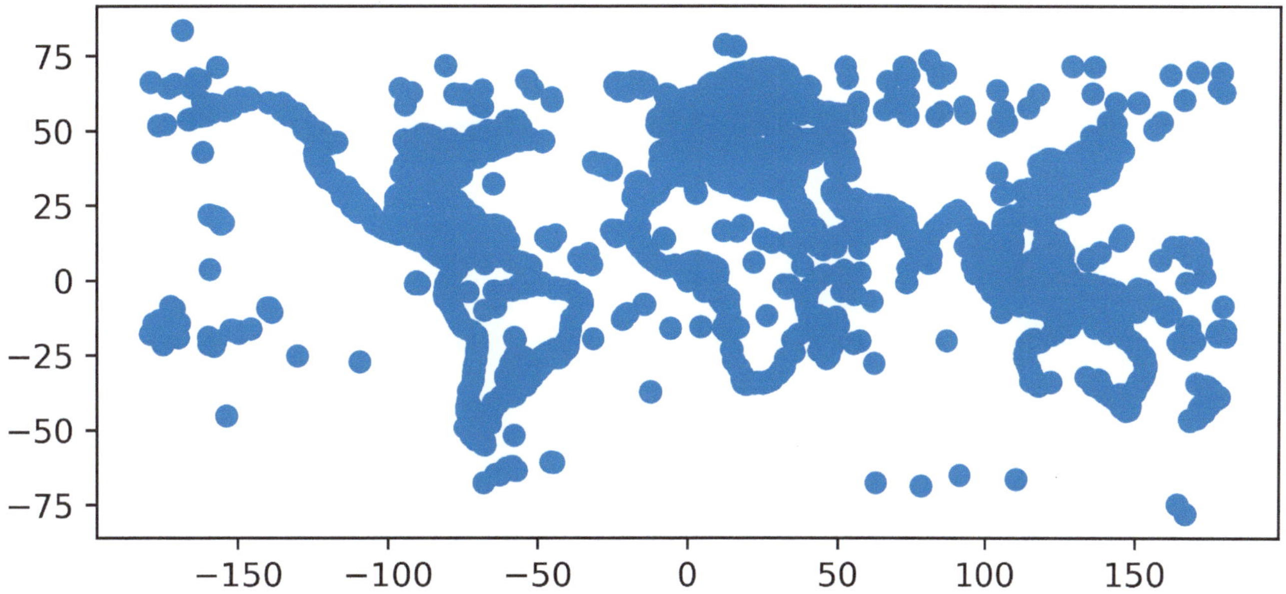

Figure 2.3: Simple plot of the World's ports.

Coordinate Reference Systems

Coordinate reference systems (CRS) are used to describe the location of geographic features on the earth's surface. They are essential for geospatial data analysis and visualization. A CRS is defined by a datum and a projection. The datum defines the shape and orientation of the earth, while the projection defines how the 3D surface of the earth is mapped onto a 2D plane.

Datum

A *datum* is a mathematical model that defines the shape, size, and orientation of the Earth's surface. The Earth is not a perfect sphere, but rather an oblate spheroid, which means that its shape is closer to a flattened sphere than a perfect sphere. Datums are needed to accurately represent the shape of the Earth's surface in geospatial data.

Datums are used in *geodesy*, which is the science of measuring and modeling the Earth's surface. They are typically represented by an ellipsoid or geoid, which is a mathematical model of the Earth's surface. An ellipsoid is a smooth surface that approximates the shape of the Earth, while a geoid is a more irregular surface that represents the Earth's gravity field.

There are two main types of datums: geographic and projected. Geographic datums are used to define the shape of the Earth's surface, while projected datums are used to map the 3D surface of the Earth onto a 2D plane.

Geographic datums are used to define the shape of the Earth's surface and are typically represented by an ellipsoid. The most commonly used geographic datum is the World Geodetic System 1984 (WGS84), which is used by the Global Positioning System (GPS) and is the standard for most geospatial data. Other common geographic datums include North American Datum 1983 (NAD83) and European Terrestrial Reference Frame 1989 (ETRF89).

Projected datums are used to map the 3D surface of the Earth onto a 2D plane. They are typically represented by a flat plane or a cylinder. Some common projected datums include Universal Transverse Mercator (UTM), which is used for large-scale mapping and navigation, and State Plane Coordinate System (SPCS), which is used for smaller-scale mapping and surveying.

Datums are needed to accurately represent the shape, size, and orientation of the Earth's surface in geospatial data. They are typically represented by an ellipsoid or geoid and are used to define the shape of the Earth's surface in geographic and projected coordinate reference systems. Common datums include WGS84, NAD83, ETRF89, UTM, and SPCS.

Projections

Projections are a method of representing the 3D surface of the Earth onto a 2D plane. Since the Earth is a 3D object and maps are 2D, projections are necessary to flatten the Earth's surface onto a map.

A datum and a projection are both important concepts in geospatial data analysis, but they serve different purposes. A datum defines the shape, size, and orientation of the Earth's surface, while a projection is a method of representing the 3D surface of the Earth onto a 2D plane. Datums are used to represent the Earth's surface in geospatial data, while projections are used to flatten the Earth's surface onto a map.

Projections are needed for several reasons. Firstly, they allow us to create maps that are easier to read and understand than a globe. Maps can provide more detailed information, such as political boundaries, transportation routes, and landmarks, that are not easily visible on a globe. Secondly, maps are useful for measuring distances, areas, and angles, which are important for a variety of applications such as surveying, navigation, and natural resource management. Finally, maps are essential for visualizing geospatial data and identifying patterns and relationships between different geographic features.

There are many different types of map projections, each with its own set of advantages and disadvantages. Some projections preserve certain properties of the Earth's surface, such as area, distance, or shape, while sacrificing others. The choice of projection depends on the purpose of the map and the region being depicted.

Here are some examples of commonly used projections:

- Mercator projection: This is a cylindrical projection that preserves the direction of true north and is often used for navigation. However, it distorts the size and shape of objects near the poles, making Greenland and Antarctica appear much larger than they actually are.
- Lambert Conformal Conic projection: This projection is commonly used for maps of the United States and Canada. It preserves the shape and size of objects near the standard parallel(s), but distorts them elsewhere on the map.
- Albers Equal Area projection: This projection preserves area, making it useful for maps that require accurate comparisons of different regions' sizes. It is commonly used for maps of the United States and other countries.
- Robinson projection: This is a compromise projection that attempts to balance the preservation of area, shape, and distance. It is a popular projection for world maps due to its attractive and well-balanced appearance.

Projections are a method of representing the 3D surface of the Earth onto a 2D plane. They are necessary for creating maps that are easier to read and understand, measuring distances and areas, and visualizing geospatial data. There are many different types of projections, each with its own advantages and disadvantages, and the choice of projection depends on the purpose of the map and the region being depicted.

Python libraries

Python has several libraries that provide functionality for working with coordinate reference systems, datums, and geodesic geometries. Here are some of the most popular libraries and how they can be used:

1. Geopandas: Geopandas is a library for working with geospatial data in Python. It provides a `GeoDataFrame` object that can store both tabular data and geometry data in one place. Geopandas uses the shapely library for working with geometries, which includes support for different types of coordinate reference systems and datums. Geopandas provides functions for transforming geometries from one coordinate reference system to another, and for calculating distances and areas using different datums.
2. Pyproj: Pyproj is a library that provides support for working with coordinate reference systems and map projections in Python. It includes a Python wrapper for the PROJ library, which is a widely used library for coordinate transformations and map projections. Pyproj provides functions for transforming coordinates between different coordinate reference systems, and for converting between different datums. It also includes support for working with different map projections, including planar projections, cylindrical projections, and conical projections.
3. Cartopy: Cartopy is a library that provides cartographic tools for Python. It includes support for working with coordinate reference systems and map projections, and provides functions for transforming coordinates between different coordinate reference systems. Cartopy also includes support for working with geodesic geometries, including calculating geodesic distances and areas. It includes support for many common map projections, including the Mercator projection and the Robinson projection.

Bringing it back to our map. The choice of projection for a plot generally depends on the area of the world that is being plotted or the purpose of the plot. Generally, when plotting data on a global scale, the Mercator or Robinson projections are used. There are two ways to do this, you can reproject the geometry column using the `to_crs` method. In this example, the crs of the `GeoDataFrame` is being reporjected from EPSG:4326 (which was set in the previous code block when creating the `GeoDataFrame`) to EPSG:3395. A number of optional kwargs have been passed to the plotting function to alter the style of the plot, e.g. markersize, color, edgecolors etc. These correspond to the styling parameters that can be passed to a matplotlib scatter plot.

```
port_geodata = port_geodata.to_crs("EPSG:3395")

fig, ax = plt.subplots()
port_geodata.plot(ax=ax, markersize=1, alpha=1, marker="*", color="green", linewidth=0.1)
plt.show()
```

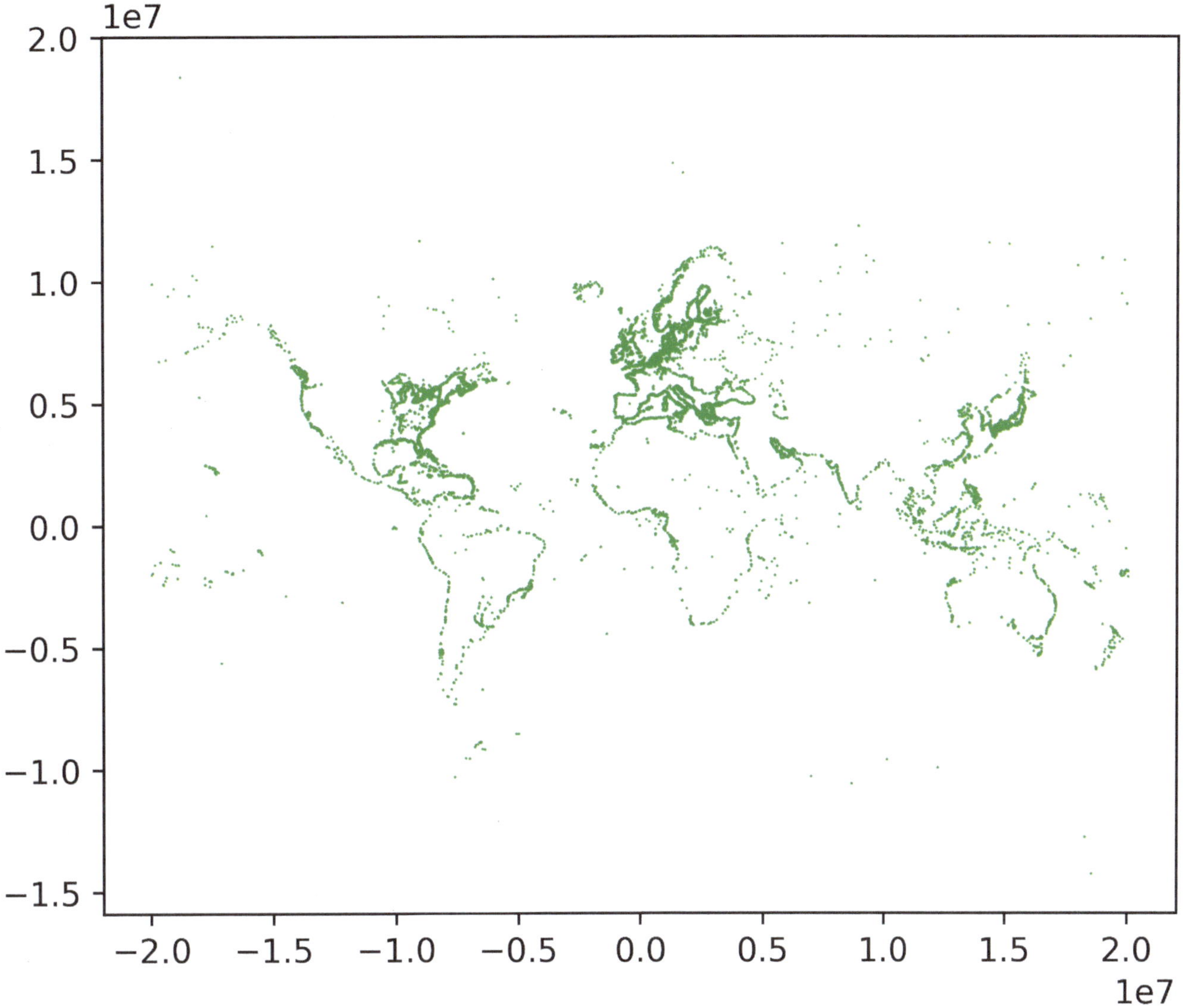

Figure 2.4: Simple plot of the World's ports in the Mercator projection.

The other option, is to use Cartopy to reproject the coordinates within the plotting operation without altering the projection within the original `GeoDataFrame`. This can be desirable if you are performing future operations on the `GeoDataFrame` and do not want to change the crs. This requires a matplotlib plotting object to be created and a projection applied to that object by cartopy. In this example, the projection is being set to the Robinson projection. Matplotlib axes and figure objects are created with `plt.subplots()` and the Robinson projection is set on the axes plotting object using the `ccrs.Robinson()` method. The transform argument is used to transform the coordinates from the GeoDataFrame crs to the Robinson projection. The `ccrs.PlateCarree()`

method is passed to transform to define (in Cartopy terms) the starting crs, which then can be transformed to the Robinson projection. The starting crs has to be defined in a way that Cartopy can understand in order for Cartopy to then reproject it.

```
import cartopy.crs as ccrs

port_geodata = port_geodata.to_crs("EPSG:4326")

fig, ax = plt.subplots(subplot_kw={'projection': ccrs.Robinson()})
port_geodata.plot(ax=ax, transform=ccrs.PlateCarree())
plt.show()
```

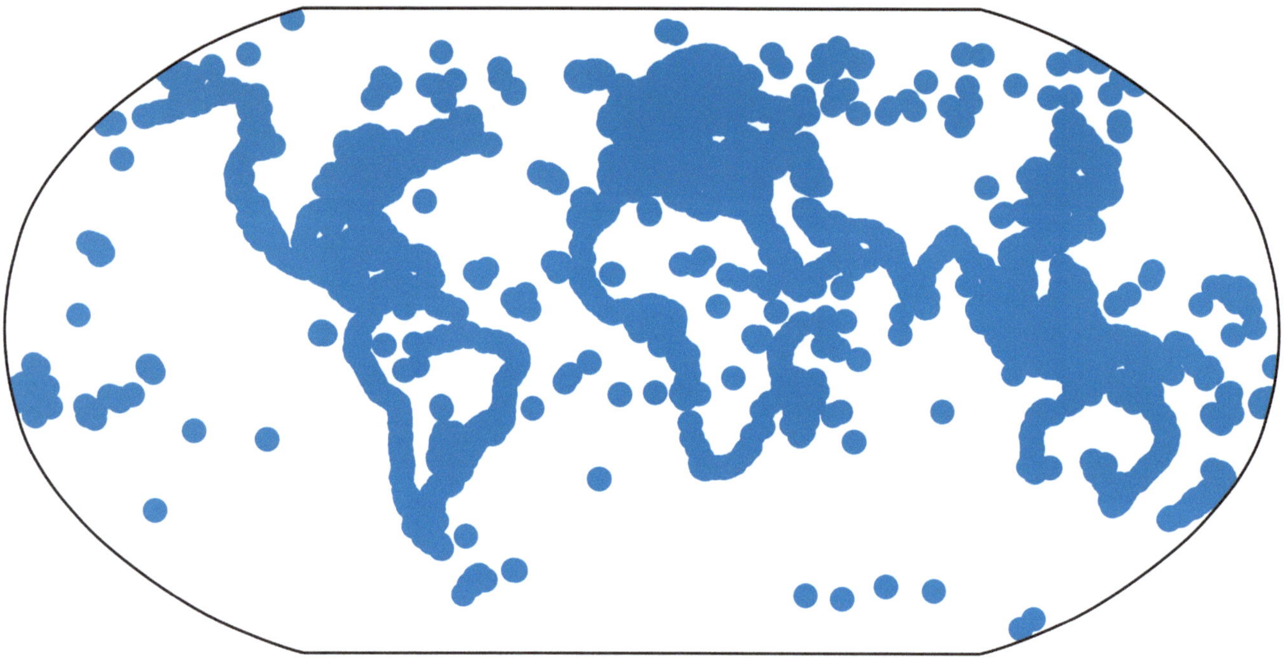

Figure 2.5: Simple plot of the World's ports in the Robinson projection.

The consequence of creating the plot within a matplotlib plotting object is that the wider matplotlib infrastructure is now easily opened up for use. In this example we are making alterations to the points as in previous examples, setting the background color to black and removing the border by setting it to the same color as the background.

```
fig, ax = plt.subplots(facecolor='black',
                       subplot_kw={'projection': ccrs.Robinson()})
ax.patch.set_facecolor('black')
port_geodata.plot(ax=ax, transform=ccrs.PlateCarree(),
             markersize=3, alpha=1, edgecolors='none', color='lightblue')
plt.setp(ax.spines.values(), color='black')
plt.setp([ax.get_xticklines(), ax.get_yticklines()], color='black')
plt.show()
```

Combining datasets within the same plot is also relatively simple. Here is an example where a csv containing the locations of the World's air-

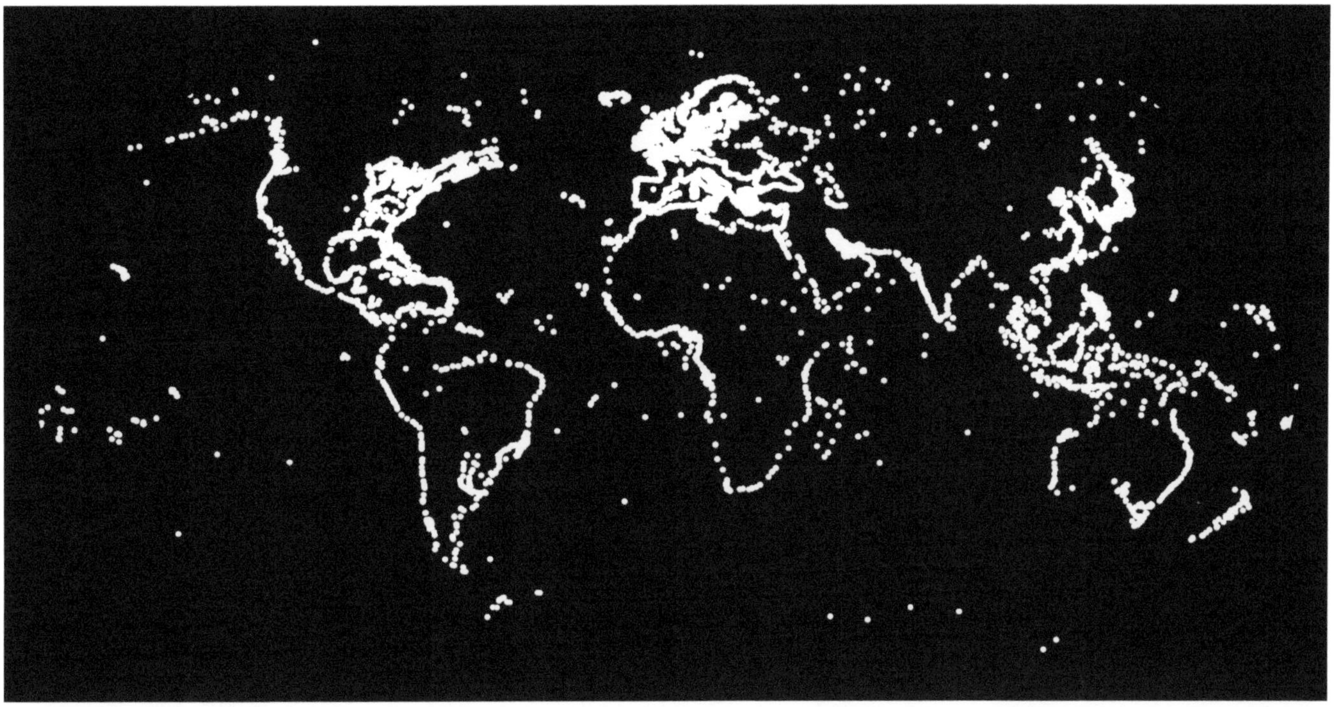

Figure 2.6: Plot of the World's ports in the Robinson projection in dark mode.

ports is loaded and converted to a GeoDataFrame in exactly the same way as the ports.

```
airports = pd.read_csv("airports", delimiter=',',
                       names=['id', 'name', 'city', 'country', 'iata',
                              'icao', 'lat', 'long', 'altitude', 'timezone',
                              'dst', 'tz', 'type', 'source'])

airport_geometry = [Point(xy) for xy in zip(airports['long'],
                                            airports['lat'])]
airport_geodata = gpd.GeoDataFrame(airports,
                                   crs="EPSG:4326",
                                   geometry=airport_geometry)
```

Plotting both datasets is simply a case of using the plot function for both GeoDataFrames and passing the same axes object to both. This tells matplotlib to plot both datasets on the same axes. In this example a number of additional parameters are used to customize the plot. Generally speaking, when plotting points, most of the *args* and *kwargs* that one would use when customizing a scatter plot can be used to customize the points plot. In this case, markersize, alpha, color and edgecolours are all tweaked.

```
fig, ax = plt.subplots(facecolor='black',
                       subplot_kw={'projection': ccrs.Robinson()})
ax.patch.set_facecolor('black')

airport_geodata.plot(ax=ax, transform=ccrs.PlateCarree(),
                     markersize=4, alpha=1, color='crimson',
                     edgecolors='none')
```

```
port_geodata.plot(ax=ax, transform=ccrs.PlateCarree(),
                  markersize=3, alpha=1, edgecolors='none', color='lightblue')

plt.setp(ax.spines.values(), color='black')
plt.setp([ax.get_xticklines(), ax.get_yticklines()], color='black')
plt.show()
```

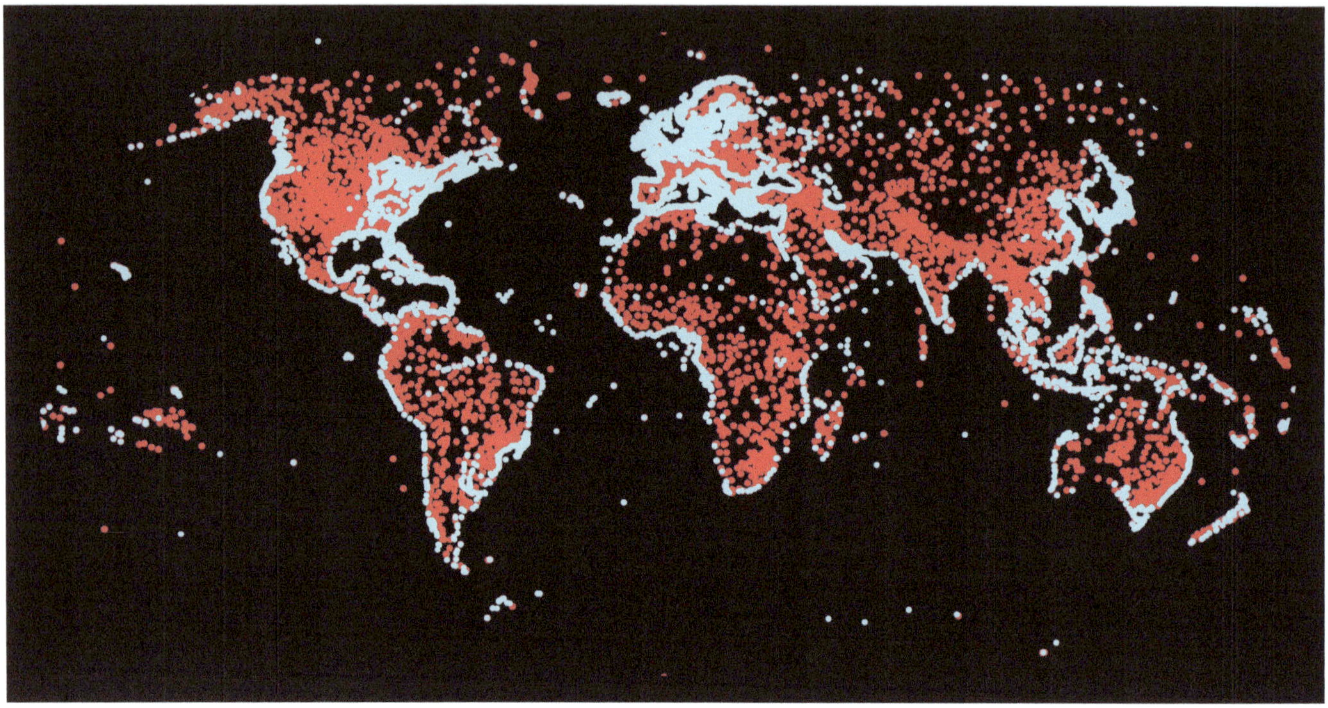

Figure 2.7: The World's airports and ports.

Hopefully, by the end of this chapter you will have an understanding of how to create, reproject and plot Points in Shapely and Geopandas. In the next chapter we will go a step further and explore LineStrings.

3

Mapping the World's Roads

3.1 Basic Line Plotting - Roads Map

Before we dive into the complexities of road network analysis, it's essential to grasp the fundamentals of linestrings in Python. In this chapter we are going to first go through an overview of linestrings in Python before moving on to a practical example where we will produce a map of the roads of France. Linestrings represent a series of connected points, forming a path. They are crucial in representing entities like roads, rivers, or any linear feature on a map. This tutorial will cover:

- Understanding what linestrings are and how they are structured in Python.
- Learning how to create linestrings using Python libraries and how to manipulate their properties.
- Techniques to visualize linestrings, enhancing your ability to interpret and present spatial data effectively.

As in the previous chapter we are going to use Shapely to construct and explore Linestrings. A LineString is a fundamental geospatial datatype that represents a sequence of connected points. It is defined by a list of ordered vertices that define the path of the LineString. In Shapely, the `LineString` class represents a linestring as a sequence of points with no area. A `LineString` can be open or closed. If the first and last point are the same, the `LineString'` is considered closed and can be used to represent features such as boundaries e.g. river basins. These are referred to as LinearRings which are a unique datatype but inherit from the `LineString` class in Python. If the first and last point are different, the `LineString` is considered open and can be used to represent features such as roads or paths. A Multilinestring is a collection of Linestrings and can be used to represent LineStrings which are linked in some way.

LineStrings

Creating a `LineString` can be accomplished in the same manner as creating a `MultiPoint` The `LineString` class is imported from the shapely geometry package and a `LineString` is created from a list of lists, with each sub-list being a x, y (longitude, latitude pair). The constructed `LineString` object represents one or more connected linear splines between the points.

```
from shapely.geometry import LineString
linestring = LineString([[2, 1], [1, 2], [1, 3], [0, 4]])
```

We can access properties as with Points. A `LineString` has zero area but a non-zero length.

```
Area 0.0
Length 3.82842712474619
Bounds (0.0, 1.0, 2.0, 4.0)
Longitude, Latitude (array('d', [2.0, 1.0, 1.0, 0.0]),
                     array('d', [1.0, 2.0, 3.0, 4.0]))
```

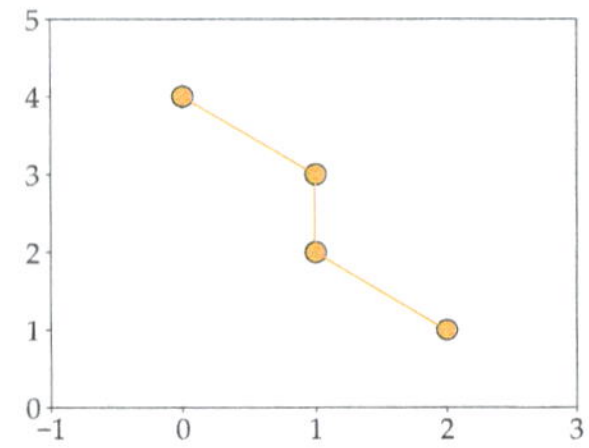

Figure 3.1: A graphical explanation of a LineString geometry.

The coordinates of LineStrings can be accessed as lists of x and y coordinates with the xy method and these can be passed to plotting functions in matplotlib to visualise the data.

```
import matplotlib.pyplot as plt


plt.plot(linestring.xy[0], linestring.xy[1], color="orange")
plt.scatter(linestring.xy[0], linestring.xy[1], s=200, color="orange",
            edgecolor="black")
plt.grid(ls="--", lw=1, alpha=0.6)
plt.xticks(np.arange(-1, 4, 1))
plt.yticks(np.arange(0, 6, 1))
plt.show()
```

It can be useful to specify position along linear features such as LineStrings and MultiLineStrings with a 1-dimensional referencing system. Shapely supports linear referencing based on length or distance, evaluating the distance along a geometric object to the projection of a given point, or the point at a given distance along the object. One useful method available for LineStrings is the interpolate function which will return a Point at a given position on the LineString. In the example below we are creating a Point halfway along the LineString.

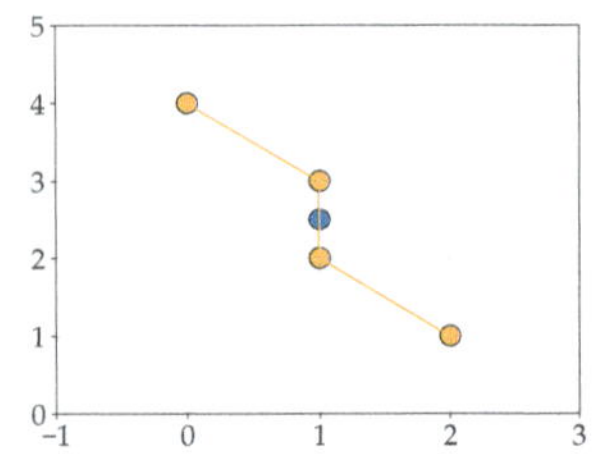

Figure 3.2: A graphical explanation of a LineString geometry with an interpolated Point.

```
middle = linestring.length / 2
interpolated_linestring = linestring.interpolate(middle)
plt.plot(linestring.xy[0], linestring.xy[1], color="orange")
plt.scatter(linestring.xy[0], linestring.xy[1], s=200, color="orange",
           edgecolor="black")
plt.scatter(interpolated_linestring.x, interpolated_linestring.y, s=200,
```

```
            edgecolor="black")
plt.grid(ls="--", lw=1, alpha=0.6)
plt.xticks(np.arange(-1, 4, 1))
plt.yticks(np.arange(0, 6, 1))
plt.show()
```

LinearRings

The Linestring point sequence may be explicitly closed by passing identical values in the first and last indices. Otherwise, the sequence will be implicitly closed by copying the first tuple to the last index. A `LinearRing` may not cross itself, and may not touch itself at a single point. Constructing a `LinearRing` is identical to constructing a `LineString`. LineStrings can be either open or closed. When a `LineString` is closed, its last point is connected to its first point, forming a loop. A `LineString` that forms a closed loop is sometimes called a `LinearRing`.

In Python, LinearRings are commonly represented as a special type of `LineString` object, where the last point is the same as the first point. LinearRings are often used to represent the boundaries of polygons, since a polygon is essentially a closed shape with straight sides.

It's worth noting that some geospatial libraries in Python, such as Shapely and GeoPandas, do not distinguish between LineStrings and LinearRings, treating them as equivalent types. Other libraries, such as gdal and Fiona, treat LinearRings as a distinct type.

```
from shapely.geometry import LinearRing
linearring = LinearRing([[-2, 3], [-2, 6], [3, 6], [3, 3]])
```

As with LineStrings, LinearRings have zero area but a non-zero length and can be plotted by accessing the x,y coordinates with the xy method.

```
Area 0.0
Length 16.0
Bounds (-2.0, 3.0, 3.0, 6.0)
Longitude, Latitude (array('d', [-2.0, -2.0, 3.0, 3.0, -2.0]),
                     array('d', [3.0, 6.0, 6.0, 3.0, 3.0]))
```

```
plt.plot(linearring.xy[0], linearring.xy[1], color="orange")
plt.scatter(linearring.xy[0], linearring.xy[1], s=200, color="orange", edgecolor="black")
plt.grid(ls="--", lw=1, alpha=0.6)
plt.xticks(np.arange(-3, 5, 1))
plt.yticks(np.arange(2, 8, 1))
plt.show()
```

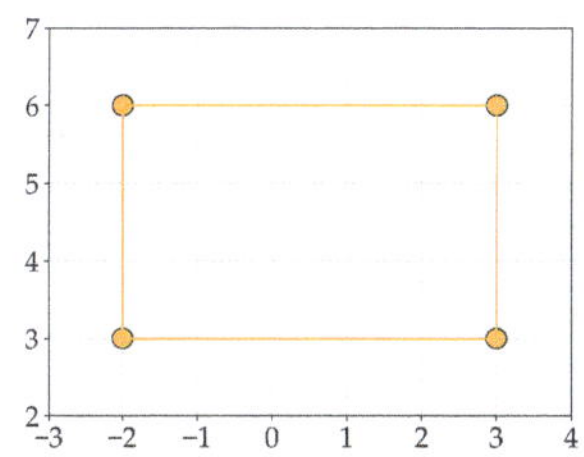

Figure 3.3: A graphical explanation of a LinearRing geometry.

MultiLineStrings

A `MultiLineString` is a collection of Linestrings. It is defined by a list of Linestring objects. In Shapely, the `MultiLineString` class represents a collection of `LineStrings` as a list of individual `LineString` objects. `MultiLineString` can be used to represent features such as a set of roads or a river system. As with Points / MultiPoints we can create a collection of LineStrings with the MultiLineString class. It is important to note that MultiLinearRings do not exist in Shapely.

```
from shapely.geometry import MultiLineString
multilinestring = MultiLineString([[[5, -1], [4, 0], [4, 2], [5, 3]],
                                   [[1, 2], [2, 3], [2, 5], [3, 6]]])
```

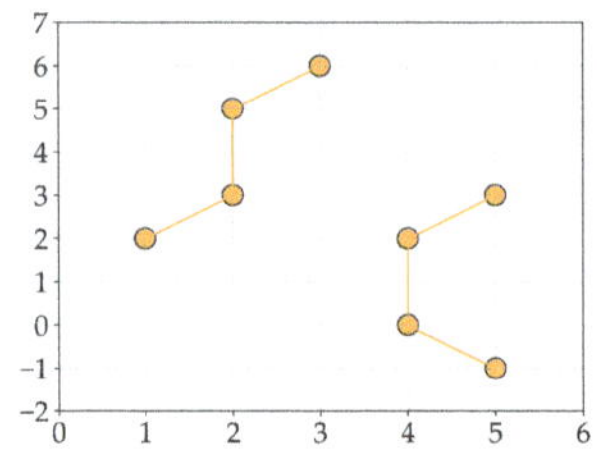

Figure 3.4: A graphical explanation of a MultiLineString geometry.

A MultiLineString has zero area and non-zero length.

```
Area 0.0
Length 9.65685424949238
Bounds (1.0, -1.0, 5.0, 6.0)
```

```
for geom in multilinestring.geoms:
    plt.plot(geom.xy[0], geom.xy[1], color="orange",)
    plt.scatter(geom.xy[0], geom.xy[1], s=200, color="orange", edgecolor="black")
plt.grid(ls="--", lw=1, alpha=0.6)
plt.xticks(np.arange(0, 7, 1))
plt.yticks(np.arange(-2, 8, 1))
plt.show()
```

GeoPandas

As with Points, Shapely is good on its own but in order to deal with large volumes of data we need to make use of Geopandas. Here we are creating a GeoDataFrame representing lines, each defined by a start and end point with associated latitude and longitude coordinates. The process begins by defining a pandas DataFrame, `lines_df`, with columns `start_lat`, `start_long`, `end_lat`, and `end_long` to represent the latitude and longitude of the start and end points of each line, respectively. Additionally, there's a lines column to serve as an identifier for each line.

The next step involves constructing the geometry list. This is done through a list comprehension that iterates over each row of `lines_df`. For each row, it creates a `LineString` object from the Shapely library. The `LineString` is formed by two `Point` objects, also from Shapely, representing the start and end coordinates of the line (extracted from `start_lat`, `start_long`, `end_lat`, and `end_long` columns).

Finally, a GeoDataFrame, `lines_gdf`, is created using GeoPandas. This `GeoDataFrame` is constructed from the original `lines_df` `DataFrame`, with the addition of the geometry column that contains the created `LineString` objects. The coordinate reference system (CRS) for lines_gdf is set to "EPSG:4326", which is the standard geographic coordinate system using latitude and longitude.

```
lines_df = pd.DataFrame(
    {
        "start_lat": [1, 9],
```

```
        "start_long": [1, 1],
        "end_lat": [9, 1],
        "end_long": [9, 9],
        "lines": [
            "a",
            "b",
        ],
    }
)
geometry = [
    LineString([Point(start_lat, start_long), Point(end_lat, end_long)])
    for start_lat, start_long, end_lat, end_long in zip(
        lines_df.start_lat, lines_df.start_long, lines_df.end_lat, lines_df.end_long
    )
]
lines_gdf = gpd.GeoDataFrame(lines_df, crs="EPSG:4326", geometry=geometry)
```

```
start_lat  start_long  end_lat  end_long lines
0          1           1        9        9     a
1          9           1        1        9     b

                                      geometry
0  LINESTRING (1.00000 1.00000, 9.00000 9.00000)
1  LINESTRING (9.00000 1.00000, 1.00000 9.00000)
```

Obviously these are all very simple examples, using a small number of Points. In the real world, most Linestrings consist of hundreds or thousands of points. In the next example, we are going to explore some truly huge datasets and use them to generate a map of the roads of France.

Mapping Roads

With a strong grasp of Linestrings under your belt, it's time to transition from theoretical concepts to practical application. In the upcoming section, we'll apply our newfound understanding of linestrings to a real-world scenario: analyzing the road network in France. This practical example will encompass:

- Applying Linestrings to Roads: Utilizing our knowledge of linestrings to interpret and analyze road data.
- Spatial Analysis of Road Networks: Exploring the road network's characteristics, focusing on how linestrings help in understanding complex spatial relationships.
- Distance Calculations and Visualization: Implementing distance calculations from a central point (Paris) and visualizing these distances using the principles learned in our linestring tutorial.

This practical application serves as a bridge between theory and real-world data, showcasing how linestrings are not just abstract concepts but powerful tools in spatial analysis. As we delve into the road network of France, the lessons from the linestring tutorial will come to life, demonstrating the practical utility and significance of these geospatial structures in understanding and visualizing our world.

We first have to load the data from a gzipped parquet file. Unlike the previous example where we used a csv file and then converted latitude and longitude values into geospatial data, here we're using `gpd.read_parquet` to read a Parquet file that already contains our geospatial data. Parquet is a highly efficient, columnar storage file format; it's designed for performance in both reading and writing complex data like geospatial data. The use of Parquet here signifies a shift to more advanced file formats optimized for handling large datasets efficiently. The line `europe = gpd.read_parquet(...)` loads the road network data for Europe from a compressed Parquet file (europe.parquet.gz). This is a large dataset, given the extensive road networks across European countries. The dataset comes from globio (Meijer, J.R., Huijbregts, M.A.J., Schotten, C.G.J. and Schipper, A.M. (2018): Global patterns of current and future road infrastructure. Environmental Research Letters, 13-064006. Data is available at www.globio.info) and has been converted from the shapefile provided at source to a parquet file prior to this exercise in order to speed up the code execution.

```
import geopandas as gpd

europe = gpd.read_parquet("europe.parquet.gz")
```

Next we need to plot the data. Some people are able to look at data and get an intuitive understanding of it, however I believe the best way to explore geospatial data is to plot it. Here we create a clear and informative map visualizing Europe's road network. This visualization serves as a foundational step in understanding the spatial characteristics of the dataset.

The code begins with creating a figure and a subplot in Matplotlib using `fig, ax = plt.subplots()`. This setup provides a canvas on which we'll draw our map. The line `europe.plot(ax=ax, lw=0.05)` plots the European roads data on the axes *ax*. The *lw=0.05* parameter sets the line width to 0.05, which is suitable for visualizing the intricate network of roads without cluttering the plot. This value can only really be determined via trial and error. In some regions of the world, much larger values are needed because the data is more sparse.

```
import matplotlib.pyplot as plt

```

```
fig, ax = plt.subplots()
europe.plot(ax=ax, lw=0.05)
ax.axis('off')
plt.show()
```

This plot gives us a good understanding of the data but it is not visually appealing. The next code snippet enhances the road network visualization by introducing thematic mapping based on the `GP_RTP` column, which corresponds to the type of road. Values of 1-5 correspond to highways, primary, secondary, tertiary and dirt tracks. The use of a dark background and a perceptually uniform colormap provides a clear, visually appealing representation of the data. Specifying a column in a GeoPandas plot tells GeoPandas to map the values in this column to a colourmap (default is *viridis*). So this map maps the values 1-5 in the *GP_RTP* column to five colours, in the `inferno_r` colourmap, meaning that each road type is assigned its own colour. This map not only highlights the road network but also conveys additional information through its color coding, offering a deeper level of insight into the dataset. With this colourmap, highways are shown at the brighter end of the colourmap and dirt tracks the darker end.

The line `fig, ax = plt.subplots(facecolor='#090909')` sets up the figure and axes with a dark background (#090909). This choice of background color adds a contemporary, sleek look to the map and enhances the visibility of brightly colored elements. The key addition here is the `column="GP_RTP"` argument in `europe.plot(...)`.

The `cmap="inferno_r"` argument applies a reversed 'inferno' color map to the plot. The 'inferno' color map is a perceptually uniform colormap, which is excellent for representing a continuous variable.

```
fig, ax = plt.subplots(facecolor='#090909')
europe.plot(ax=ax, column="GP_RTP", cmap="inferno_r", lw=0.05)
ax.axis('off')
ax.set_xlim(-12.5, 30)
ax.set_ylim(35, 73)
plt.show()
```

Spatial Joins in GeoPandas

We are going to pause here to briefly explain how spatial joins work in Python. The next step is to reduce our dataset down to only include the roads of France. To do this we are going to perform a spatial join on the roads data to find all roads that are within a polygon representing France.

The `sjoin` method in GeoPandas is a cornerstone function for spatial

Figure 3.5: Roads of Europe.

analysis, enabling the joining of two GeoDataFrames based on their spatial relationship. This method relies on various types of spatial re-

Figure 3.6: Roads of Europe with lines coloured according to their type.

lationships, each with distinct characteristics and applications. Let's delve into these relationships to understand how they are employed in spatial joins.

1. **Contains**
   Definition: A geometry A contains another geometry B if all points of B lie within A, and at least one point of the interior of B lies in the interior of A. Notably, A does not contain B if any point of B lies on the boundary of A. **Applicability**: This is used in scenarios where you need to check if one geometry is completely within another without touching its boundary. Possible Combinations: It applies to various combinations like Points within LineStrings, LineStrings within Polygons, etc.
2. **Within**
   Definition: A geometry A is within another geometry B if all points of A lie within B. This is the inverse of contains. The boundary of B is not considered within B. Inverse Relationship: Within is the inverse of contains. Thus, Within(A, B) is equivalent to Contains(B, A). **Applicability**: Used to find geometries completely enclosed by others. It's crucial in scenarios like finding locations entirely within a specific area.
3. **Crosses**
   Definition: A geometry A crosses another geometry B if their interiors intersect but do not completely contain each other, and the intersection has a lower dimension than the geometries themselves. **Applicability**: It's used to find geometries that intersect in a non-contained manner, like roads crossing each other.
4. **Intersects**
   Definition: Geometries A and B intersect if they share any portion of space. This is a broader relationship than crosses or contains. **Applicability**: This relationship is used in a wide range of scenarios, like finding any overlapping or touching features.
5. **Overlaps**
   Definition: A geometry A overlaps B when they share some part of space but are not completely contained within each other, and they are of the same dimension. **Applicability**: It's typically used to find geometries that are partially, but not entirely, covering each other, like overlapping regions.
6. **Touches**
   Definition: Geometries A and B touch if they have at least one point in common, but their interiors do not intersect. **Applicability**: This relationship is significant for identifying adjacent features, like countries sharing a border.

When performing a spatial join using the `sjoin` method in GeoPandas, these spatial relationships determine how the datasets are combined. The choice of relationship depends on the specific nature of the spatial query or analysis being conducted. For example:

Practically, we first load the data for France and then perform the join. We are using the ne_10m_admin_0_countries dataset from NaturalEarth but any shapefile with Polygons representing the world will work. `gpd.read_file(...)` loads a shapefile containing country boundaries into a GeoDataFrame df and `df.loc[df['ADMIN'] == 'France']` filters to get a new `GeoDataFrame` called `gdf` containing only the data for France. We use the ADMIN column, which holds country names, to filter down to France.

`gpd.sjoin(europe, gdf, predicate='within')` joins the European roads dataset with the France dataset based on the roads being within France. It filters out all road segments that aren't completely inside France's borders.

There are a few things to consider. For these spatial relationships to be valid, the source geometries must have the same Spatial Reference Identifier (SRID), meaning they should be in the same coordinate projection. It is also important to understand how these relationships work in geospatial analysis, as it affects the outcomes of spatial joins and queries.

```
df = gpd.read_file('NaturalEarth/10m_cultural/ne_10m_admin_0_countries.shp')
gdf = df.loc[df['ADMIN'] == 'France']

roads = gpd.sjoin(europe, gdf, op='within')
```

As above, we are going to plot the data to get a sense for what it looks like. The subtle adjustment in line width enhances the visual clarity of the roads as we are now plotting a smaller number of them. The only real difference here is that we have increased the linewidth slightly.

```
import matplotlib.pyplot as plt

fig, ax = plt.subplots()
roads.plot(ax=ax, lw=0.25)
ax.axis('off')
plt.show()
```

This map is already pretty good and can be customised from here to produce a map in dark mode or with different colours for different road types as we did above. However, to add some additional complexity, we are going to calculate the distance of each section of the road network in France from Paris, the nation's capital. This analysis can reveal patterns of centrality and peripherality in the road network, offering a clearer picture of how different regions are connected to the capital. Using GeoPandas we'll calculate the distance from each road segment in our dataset to Paris. The calculated distances will then be visualized, providing a graphical representation of how the road network spans out from Paris.

We start by defining the geographical coordinates of Paris. This serves as our reference point for all subsequent distance calculations. The `Point` class from the Shapely library is used to create a point geometry. Here, `Point(2.3522, 48.8566)` creates a point representing the geographical coordinates of Paris (longitude 2.3522, latitude 48.8566).

`gpd.GeoDataFrame({'geometry': [Point(...)]}, crs="EPSG:4326")`, creates a `GeoDataFrame` with one column named 'geometry'. This column contains our Shapely Point object, representing Paris.
The `crs="EPSG:4326"` specifies the coordinate reference system of the `GeoDataFrame`. EPSG:4326 is a widely used geographic coordinate system (GCS) based on the WGS84 datum.

```
import geopandas as gpd
from shapely.geometry import Point

gdf = gpd.GeoDataFrame({'geometry': [Point(2.3522, 48.8566)]},
                       crs="EPSG:4326")
```

We can then use this GeoDataFrame to calculate the distance from our roads to Paris. The distance method in GeoPandas (which is the distance method in Shapely, applied to all geometries in the geometry column) calculates the distance between each geometry in the roads `GeoDataFrame` and a single geometry, in this case, the `Point` geometry representing Paris from our `gdf`. `gdf.iloc[0].geometry` extracts the geometry of the first (and only) row in our gdf, which is the `Point` representing Paris and `roads.distance(...)` applies the distance calculation between each road segment in roads and the Point representing Paris. `roads['distance2paris'] = distances` adds a new column, *distance2paris*, to the roads `GeoDataFrame`. This column stores the computed distances, effectively tagging each road segment with its respective distance from Paris.

```
import geopandas as gpd
from shapely.geometry import Point, LineString

distances = roads.distance(gdf.iloc[0].geometry)
roads['distance2paris'] = distances

print(roads)
```

Finally, we want to scale the size of our lines depending on the distance to Paris. The following mathematical transformations and normalization is an example of an Exponential Decay Function.

`roads['distance2paris_lw'] = 1 / np.exp(roads['distance2paris'])` applies an exponential decay function to the distance values. This transformation is useful for diminishing the effect of large distance values, which can be particularly helpful in visualization and clustering tasks where relative distance is more important than absolute distance. *leftSpan* is calculated as the range of the transformed distance values (distance2paris_lw). *rightSpan* defines the desired range of the normalized values. In this case, it's set between 0.05 and 0.9. `valueScaled = (roads['distance2paris_lw'] - np.amin(roads['distance2paris_lw'])) / leftSpan` scales the transformed values to a 0-1 range. The final line `roads['distance2paris_lw'] = 0.05 + (valueScaled * rightSpan)` shifts and scales these values to the desired range (0.05 to 0.9). This normalization ensures that the transformed distances are mapped to a consistent range, making them easier to compare and visualize.

```
import numpy as np

roads['distance2paris_lw'] = 1 / np.exp(roads['distance2paris'])

leftSpan = np.amax(roads['distance2paris_lw']) - np.amin(roads['distance2paris_lw'])
rightSpan = 0.9 - 0.05
valueScaled = (roads['distance2paris_lw']
               - np.amin(roads['distance2paris_lw'])) / leftSpan
roads['distance2paris_lw'] = 0.05 + (valueScaled * rightSpan)
```

Finally, we can plot the data as we have in previous examples but use our `distance2paris_lw` column to scale the size of the roads by explicitly setting the column values as the plot linewidth width: `lw=roads['distance2paris_lw']`

```
fig, ax = plt.subplots(facecolor='#090909')

roads.plot(ax=ax, column="GP_RTP", cmap="afmhot_r", lw=roads['distance2paris_lw'])
ax.axis('off')

plt.show()
```

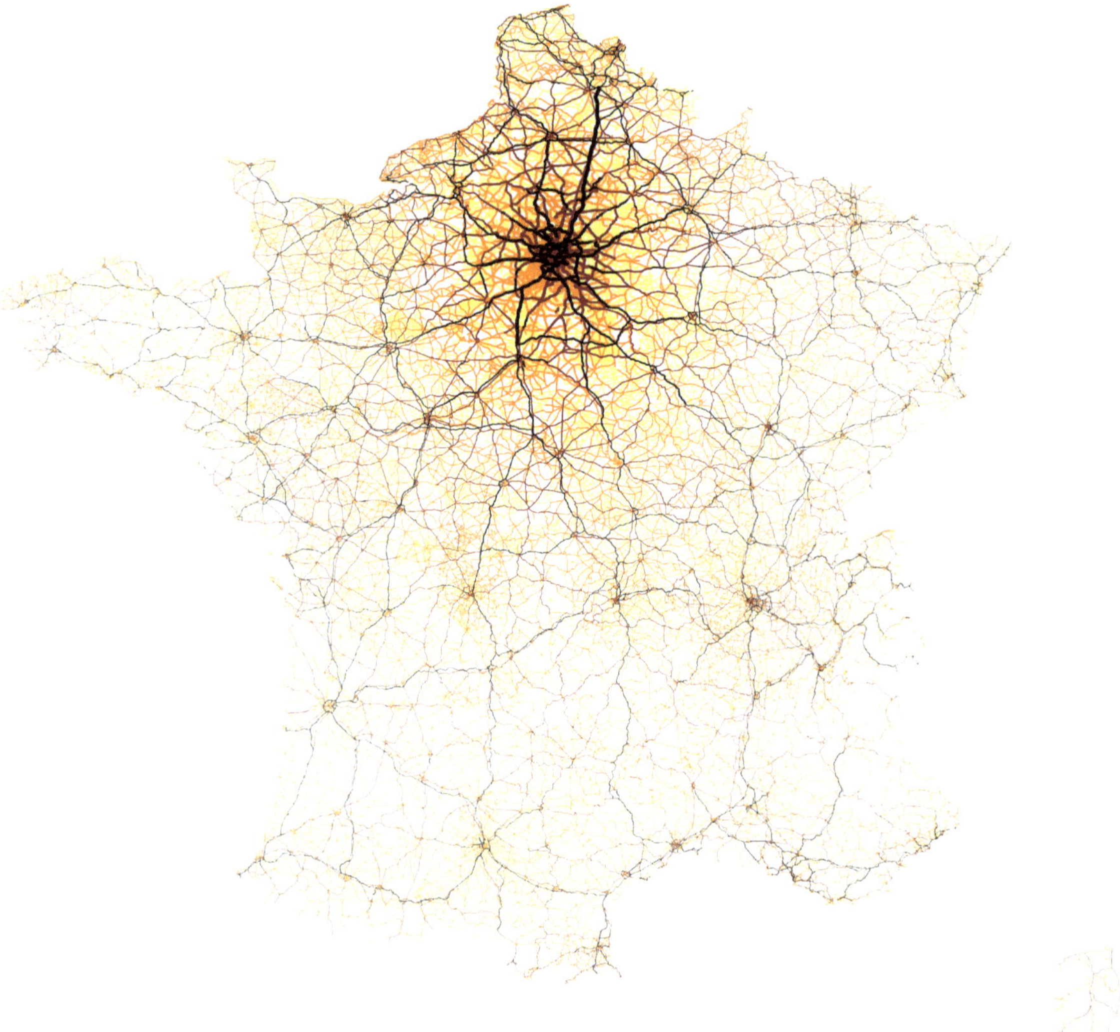

Figure 3.7: Roads of France with lines scaled by their distance to Paris.

4

Mapping Historical Empires

4.1 Polygons

In this section we are going to go through an overview of Polygons in Python, and then apply that background to a practical example: mapping the territories of different empires.

Before we delve into the historical expanse of empires, let's lay the groundwork with a fundamental understanding of polygons in Python. In geospatial analysis, polygons are key to representing areas such as countries, lakes, or other landmasses. This tutorial aims to provide you with a comprehensive introduction to polygons, essential for any geospatial data enthusiast or professional.

In this tutorial, we will cover the following key aspects:

- Defining what polygons are in the context of geospatial data and their significance in representing geographical areas.
- We'll explore how to create polygons using Python libraries, manipulate their properties, and understand their attributes.
- Practical tips and techniques will be provided for effectively visualizing polygons. This is crucial for both data interpretation and presentation.

A `Polygon` is a fundamental geospatial datatype that represents a closed shape with straight sides. It is defined by a set of ordered vertices that define the boundary of the polygon. In Shapely, the `Polygon` class represents a polygon as a closed ring of points with the first and last point being the same. The polygon can have holes in its interior, which are represented as other closed rings of points. Polygons are often used to represent features such as countries, states, or neighborhoods.

A `MultiPolygon` is a collection of Polygons. It is defined by a list of `Polygon` objects. In Shapely, the `MultiPolygon` class represents a collection of Polygons as a list of individual Polygon objects. MultiPolygons can be used to represent features such as a set of countries or a collection of neighborhoods.

Polygons

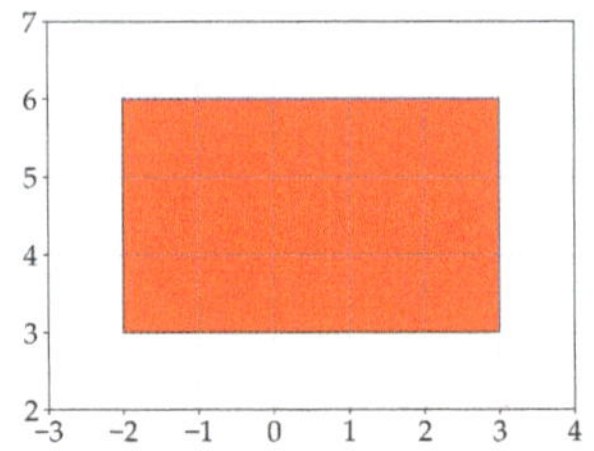

Figure 4.1: A graphical explanation of a Polygon geometry.

Polygons can be created by importing the Polygon class from the shapely geometry module and passing a list of lists, with the sublist corresponding to points defining the boundary of the object.

```
from shapely.geometry import Polygon
polygon = Polygon([[-2, 3], [-2, 6], [3, 6], [3, 3]])
```

A `Polygon` has non-zero area and non-zero length. Hopefully the relationship between Points, Lines and Polygons is now becoming clear. The boundary of a `Polygon` is a `LineString` and the center of a `Polygon` is a `Point`. The Centroid is a representation of the object's geometric centroid (point).

```
print("Area", polygon.area)
print("Length", polygon.length)
print("Bounds", polygon.bounds)
print("Boundary", polygon.boundary)
print("Center", polygon.centroid)
```

```
Area 15.0
Length 16.0
Bounds (-2.0, 3.0, 3.0, 6.0)
Boundary LINESTRING (-2 3, -2 6, 3 6, 3 3, -2 3)
Center POINT (0.5 4.5)
```

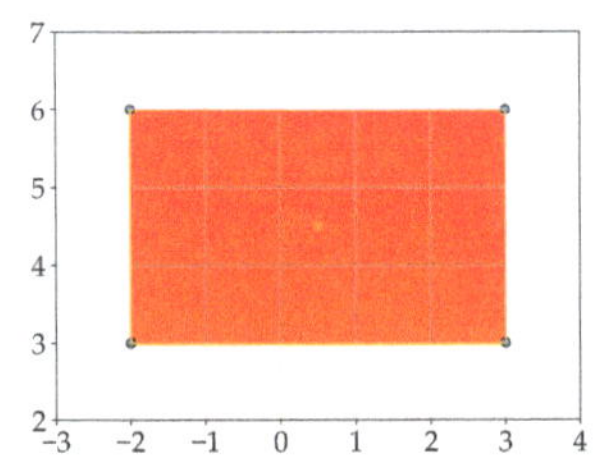

Figure 4.2: A graphical explanation of a Polygon geometry showing the boundary Points (Blue), Centroid (Orange).

Plotting a polygon is similar to that of a `LineString` and we can also plot the boundary, center etc.

```
import matplotlib.pyplot as plt

plt.fill(*polygon.exterior.xy, color="red", edgecolor="black")
plt.scatter(polygon.exterior.xy[0], polygon.exterior.xy[1],edgecolor="black")
plt.plot(polygon.boundary.xy[0], polygon.boundary.xy[1], color="orange")
plt.scatter(polygon.centroid.x, polygon.centroid.y)
plt.grid(ls="--", lw=1, alpha=0.6)
plt.xticks(np.arange(-3, 5, 1))
plt.yticks(np.arange(2, 8, 1))
plt.show()
```

MultiPolygons

Collections of Polygons are called MultiPolygons. These are probably the most common collection that you'll find in the wild. Country shapefiles are often described by MultiPolygons, e.g. The United Kingdom (if we exclude all of the islands) can be described by a `MultiPolygon` containing a `Polygon` for mainland Great Britain and a Polygon for Northern Ireland, The USA would be the same with Polygons for the mainland, Alaska etc.

In this code snippet, we are using it to create a `MultiPolygon`, which is a geometric object consisting of multiple polygons. Here, we are creating a `MultiPolygon` object consisting of two polygons. `Polygon([[4, 4], [4, 6], [7, 6], [7, 4]])` creates the first polygon. The points `[[4, 4], [4, 6], [7, 6], [7, 4]]` represent the vertices of the polygon in the (x, y) coordinate system. `Polygon([[4, 0], [4, 3], [7, 3], [7, 0]])` creates the second polygon with a different set of vertices. Each `Polygon` object is defined by a list of tuples, where each tuple represents a vertex of the polygon. The variable *multipolygon* now holds a `MultiPolygon` object which contains two separate polygons as defined by the points.

```
from shapely.geometry import MultiPolygon, Polygon
multipolygon = MultiPolygon([Polygon([[4, 4], [4, 6], [7, 6], [7, 4]]),
                             Polygon([[4, 0], [4, 3], [7, 3], [7, 0]])])
```

We can also calculate a number of properties. The area property calculates the total surface area enclosed by all the polygons within the `MultiPolygon`. This is particularly useful in spatial analysis for measuring the size of a geographic feature. The length property measures the total perimeter of all the polygons in the `MultiPolygon`. This includes the length of their outer and inner boundaries. It's used for calculating the boundary lengths of geographic features. The bounds property provides the coordinates of the smallest rectangle that can completely enclose the `MultiPolygon`. It returns a tuple of the form (min_x, min_y, max_x, max_y), representing the coordinates of the bounding box. The boundary property returns a geometric object that represents the combined boundaries of all the polygons in the `MultiPolygon`. It's useful for understanding the limits or outlines of a complex spatial feature. The centroid property calculates the centroid, or the geometric center, of the `MultiPolygon`. It is the point where the `MultiPolygon` would balance if it were made of a material with uniform density. This is key in applications where a central or representative point of a region is needed.

```
print("Area", multipolygon.area)
print("Length", multipolygon.length)
print("Bounds", multipolygon.bounds)
print("Boundary", multipolygon.boundary)
print("Center", multipolygon.centroid)
```

```
Area 15.0
Length 22.0
```

```
Bounds (4.0, 0.0, 7.0, 6.0)
Boundary MULTILINESTRING ((4 4, 4 6, 7 6, 7 4, 4 4), (4 0, 4 3, 7 3, 7 0, 4 0))
Center POINT (5.5 2.9)
```

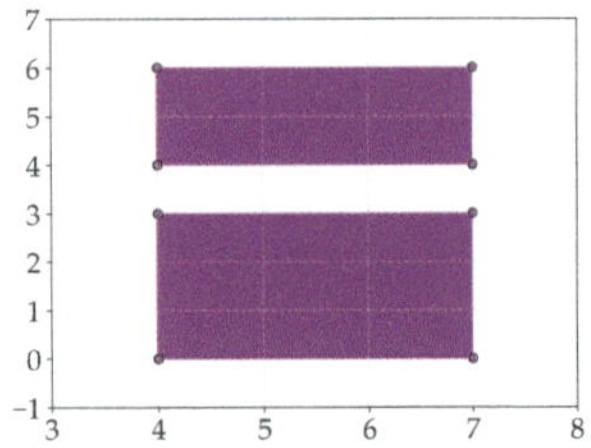

Figure 4.3: A graphical explanation of a MultiPolygon geometry.

Finally, we can also plot the `MultiPolygon` in the same manner as with our regular polygons.

```
for geom in multipolygon.geoms:
    plt.fill(*geom.exterior.xy, color="purple", label="MultiPolygon")
    plt.scatter(geom.exterior.xy[0], geom.exterior.xy[1], color="purple",
                edgecolor="black")
plt.grid(ls="--", lw=1, alpha=0.6)
plt.xticks(np.arange(3, 9, 1))
plt.yticks(np.arange(-1, 8, 1))
plt.show()
```

Geopandas

As in the previous two examples, Shapely cannot be used on its own and needs to be combined with Geopandas. Here we are creating a `GeoDataFrame` that represents polygons, each originating from a point with associated latitude and longitude coordinates, and then expanding these points into circular areas. The process begins with the creation of a pandas `DataFrame`, *polygon_df*, which contains three columns: lat and long for the latitude and longitude of the center points of the polygons, and polygons as an identifier for each polygon.

Next, the geometry list is constructed using a list comprehension. For each pair of latitude (lat) and longitude (long) values from *polygon_df*, a `Point` object (from the Shapely library) is created. This results in a list of Point objects, each representing the center of a polygon.

A `GeoDataFrame`, *polygon_gdf*, is then created from this using GeoPandas. This `GeoDataFrame` is initially similar to *polygon_df*, but with an added geometry column containing the `Point` objects. The coordinate reference system (CRS) for *polygon_gdf* is set to "EPSG:4326", a standard geographic coordinate system using latitude and longitude.

The final line of the code transforms these points into polygons. This transformation is done by applying the buffer method to each point in the geometry column of *polygon_gdf*. The `buffer(1.0)` method creates a circular polygon (buffer) around each point with a radius of 1.0 units (in the coordinate system of the `GeoDataFrame`, typically degrees in EPSG:4326). The result is a `GeoDataFrame` where each row represents a circular area (a polygon) with a specific radius, centered around the original point coordinates.

```
print("Area", multipolygon.area)
```

```
print("Length", multipolygon.length)
print("Bounds", multipolygon.bounds)
print("Boundary", multipolygon.boundary)
print("Center", multipolygon.centroid)
```

Mapping Empires

Having grasped the essentials of polygons, it's time to apply this knowledge to a historical and practical context: the analysis of empires. This practical application will include:

- Utilizing our understanding of polygons to map and analyze the geographical boundaries of various empires.
- Exploring the rise and fall of empires through the lens of geospatial data, observing how territorial changes can be represented and analyzed using polygons.
- Implementing visualization techniques to bring the historical data to life, making it easier to interpret and understand the historical significance of these empires.

This requires the use of some geosjson files (`https://github.com/symmy596/historical-geospatial-data`), specifically a geojson of the Roman and Mongolian empires at their respective peaks. These two lines of code are used to load geographical data of the Roman and Mongol empires from their respective GeoJSON files, creating two GeoDataFrames ready for spatial analysis or visualization.

`import geopandas as gpd` imports the GeoPandas library and aliases it as gpd.

`gpd.read_file(...)` reads the GeoJSON files named "roman_empire.geojson" and "mongol_empire.geojson". These files contain the spatial data for the Roman and Mongol Empires. The read_file method loads this data and stores it in a GeoDataFrame named roman_empire. Important to note that this is a new filetype and while we used read_parquet in the previous chapter, we are using `read_file‘` here. `read_file‘` is appropriate for a number of filetypes, namely geojson files and shapefiles.

```
import geopandas as gpd

roman_empire = gpd.read_file("roman_empire.geojson")
mongol_empire = gpd.read_file("mongol_empire.geojson")
```

The following code is used to load the NaturalEarth country ADMIN boundaries. We are going to use these as our base layer so that we can see the modern day countries that these empires span. This is a common data layer that we will employ throughout this tutorial. One thing to note is that we are removing Antarctica and converting the country Polygons into LineStrings using the boundary method.

```
df = gpd.read_file('NaturalEarth/10m_cultural/ne_10m_admin_0_countries.shp')
df = df[df['CONTINENT'] != 'Antarctica']
df['boundary'] = df.geometry.boundary
df['geometry'] = df['boundary'].copy()
```

The `plt.subplots` function is used to create a figure (*fig*) and a subplot (*ax*). The *subplot_kw* argument specifies that the subplot uses a Mercator projection from Cartopy (`ccrs.Mercator()`), ideal for mapping purposes. The `figsize=(10,10)` sets the size of the figure. `df.plot(...)` plots another `DataFrame` df on the map with a linewidth of 0.1. The `transform=ccrs.PlateCarree()` indicates the coordinate system of the data. `roman_empire.plot(...)`, plots the Roman Empire data in crimson color and `mongol_empire.plot(...)`, plots the Mongol Empire data in blue colors.

```
import matplotlib.pyplot as plt
import cartopy.crs as ccrs

fig, ax = plt.subplots(subplot_kw={'projection': ccrs.Mercator()},
                       figsize=(10,10))
df.plot(ax=ax, transform=ccrs.PlateCarree(), color='black', lw=0.1)

roman_empire.plot(ax=ax,
                  transform=ccrs.PlateCarree(),
                  color='crimson')
mongol_empire.plot(ax=ax,
                   transform=ccrs.PlateCarree(),
                   color='blue')

ax.axis('off')
plt.show()
```

Spatial Relationships

We are going to isolate the parts of these two empires that overlap, to showcase the use of spatial relationships in GeoPandas. Spatial relationships are fundamental concepts in geospatial analysis, allowing us to understand how different geographic features interact with each other in space. In GeoPandas, these relationships are explored through various methods. These methods, including `union`, `intersection`, `symmetric_difference`, and `difference`, enable the analysis and manipulation of geometric shapes, facilitating a deeper understanding of their spatial interactions. Here's a closer look at each of these key methods:

- **Union**: The `union` method is used to combine the spatial information of two GeoSeries or GeoDataFrames. It creates a new geometry that represents the spatial set union of the geometries

Figure 4.4: Mongol (Bue) and Roman (Red) Empires

of the two inputs. This method is particularly useful when you want to merge overlapping or adjacent geographic areas into a single area.

- **Intersection**: The `intersection` method computes the shared or overlapping part of two geometries. When applied to two GeoSeries or GeoDataFrames, it returns a new geometry representing the common spatial area between the input geometries. This is often used in spatial analysis to find common features between different spatial layers, such as finding the overlapping regions of two land use types.

- **Symmetric Difference**: The `symmetric_difference` method returns a geometry that represents the spatial difference of two input geometries, excluding the intersection. Essentially, it produces a new geometry covering the area which lies in either of the input geometries but not in their intersection. This method is useful for understanding the spatial characteristics unique to each of the input geometries.
- **Difference**: The `difference` method is used to compute the difference between two geometries. It returns a geometry representing the part of the first (caller) geometry that does not intersect with the second geometry. This is particularly useful for subtracting one geographic layer from another, such as removing a water body from a land use map to focus only on the land regions.

These spatial operations provide insightful ways to analyze and visualize the geographical relationships between the Roman and Mongol empires. Whether it's understanding their combined reach (`union`), their distinct territories (`difference` and `symmetric_difference`), or their overlapping regions (`intersection`), each operation serves a unique purpose in geospatial analysis.

`roman_empire.overlay(mongol_empire, how='union')` creates a new `GeoDataFrame` union. The union operation combines the spatial extents of both the Roman and Mongol empires. It includes all areas covered by either empire, essentially merging their geographic footprints.

`roman_empire.overlay(mongol_empire, how='difference')` generates a
`GeoDataFrame` difference showing the area covered by the Roman Empire but not by the Mongol Empire. It essentially subtracts the Mongol Empire's area from the Roman Empire's.

`roman_empire.overlay(mongol_empire, how='symmetric_difference')` results in a `GeoDataFrame` `symmetric_difference`, which includes areas that are in either the Roman or Mongol empire, but not in both. It's a way to visualize the unique areas of each empire, excluding their overlap.

`roman_empire.overlay(mongol_empire, how='intersection')` creates a
`GeoDataFrame` intersection showing the geographical area where both the Roman and Mongol empires overlapped.

```
union = roman_empire.overlay(mongol_empire, how='union')
difference = roman_empire.overlay(mongol_empire, how='difference')
symmetric_difference = roman_empire.overlay(mongol_empire, how='symmetric_difference')
intersection = roman_empire.overlay(mongol_empire, how='intersection')
```

This code snippet highlights the intersection (overlapping regions) of the Roman and Mongol empires against a broader geospatial context provided by df. `intersection.plot(...)` the intersection `GeoDataFrame`, which contains the overlapping areas of the Roman and Mongol empires, is plotted. It's transformed to the PlateCarree projection and colored purple.

```
import matplotlib.pyplot as plt
import cartopy.crs as ccrs

fig, ax = plt.subplots(subplot_kw={'projection': ccrs.Mercator()},
                       figsize=(10,10))

df.plot(ax=ax, transform=ccrs.PlateCarree(), color='black', lw=0.1)

intersection.plot(ax=ax,
             transform=ccrs.PlateCarree(),
             color='purple')

ax.set_ylim(0, 9000000)
ax.set_xlim(-100000, 14000000)
ax.axis('off')
plt.show()
```

This creates a layered geospatial visualization. By plotting the Roman and Mongol empires, and their intersection, it allows us to visually explore their historical extents and overlaps. The use of distinct colors for each empire and their intersection area enhances the clarity and educational value of the map.

`df.plot(...)` plots underlying modern day borders with a black line and a 0.1 linewidth, `roman_empire.plot(...)` plots the Roman Empire in a crimson color. `mongol_empire.plot(...)` plots the Mongol Empire in a blue. Finally, `intersection.plot(...)` plots the intersection of the Roman and Mongol empires in purple. This emphasizes the areas where the two empires overlapped geographically.

```
import matplotlib.pyplot as plt
import cartopy.crs as ccrs

fig, ax = plt.subplots(subplot_kw={'projection': ccrs.Mercator()},
                       figsize=(10,10))
df.plot(ax=ax, transform=ccrs.PlateCarree(), color='black', lw=0.1)
```

Figure 4.5: Overlapping areas of the Mongol and Roman empires.

```

roman_empire.plot(ax=ax,
                  transform=ccrs.PlateCarree(),
                  color='crimson')
mongol_empire.plot(ax=ax,
                   transform=ccrs.PlateCarree(),
                   color='blue')
intersection.plot(ax=ax,
                  transform=ccrs.PlateCarree(),
```

```
                color='purple')

ax.axis('off')
plt.show()
```

Figure 4.6: Mongol (Bue) and Roman (Red) Empires with overlapping regions shown in Purple.

5

Mapping Livestock Populations

5.1 Basic Raster Plotting Livestock Map

In this chapter we are going to go through a simple example to lay the foundation for raster plotting in Python. We are going to take a raster showing the population density of cattle around the world and plot it using Python's *Rasterio library*. This section is designed to provide the fundamentals needed for some of the more complicated examples to be discussed in later chapters.

In the world of geospatial analysis, raster data plays a pivotal role in representing and interpreting the Earth's surface. As we embark on this section of our journey, we're focusing on the fundamental aspects of raster data – from its nature and structure to its various applications in geospatial science. Our goal is to provide you with a thorough understanding of raster data and equip you with the skills to manipulate and visualize it effectively using Python's rasterio library.

Raster data, is essentially a grid of cells or pixels and captures various attributes of the Earth's surface, such as elevation, temperature, or land cover. Each pixel in this grid holds a value corresponding to the attribute it represents, making raster data a powerful tool for spatial analysis. Raster data finds applications in remote sensing, land use mapping, environmental monitoring, and climate change analysis. It's instrumental in classifying land cover, tracking environmental changes, and studying the impacts of climate change. The resolution of raster data, indicating the smallest discernible feature, plays a critical role in the level of detail and accuracy of the data.

Raster Data

There are several raster formats used in geospatial data. Here are some of the most common ones:

- **GeoTIFF**: GeoTIFF is a widely used raster format that allows for the embedding of georeferencing information within the raster file. This format is widely supported by many software packages and contains metadata that describes the spatial reference system, resolution, and other information about the data.
- **JPEG/JPEG2000**: JPEG and JPEG2000 are commonly used image formats that can be used for raster data. They are lossy compression formats that allow for efficient storage and transmission of large amounts of data. However, due to their compression, they may not be suitable for some types of analysis.
- **PNG**: PNG is a lossless image format that is commonly used for displaying and sharing raster data. It can store alpha channel information and supports transparency.
- **ESRI Grid**: ESRI Grid is a proprietary raster format developed by ESRI. It is commonly used in the ArcGIS software package and can store a variety of data types including integer, float, and double. It also supports the embedding of georeferencing information within the file.
- **NetCDF**: NetCDF is a multidimensional data format commonly used in scientific applications. It can store raster data with metadata that describes the spatial and temporal information associated with the data.
- **HDF**: HDF is a data format developed by the National Center for Supercomputing Applications (NCSA). It is commonly used in scientific and remote sensing applications and can store a variety of data types including raster data. It also supports the embedding of georeferencing information within the file.

Each raster format has its own characteristics and advantages depending on the use case. It is important to choose the appropriate format based on the analysis requirements, data size, and software compatibility.

Raster operations

Raster operations are fundamental in processing raster data, and they involve a wide range of operations that modify or transform raster data. Here are the different types of raster operations:

- **Reclassification**: This is the process of modifying the values of raster cells based on specific rules. It involves assigning new values to the raster cells based on a set of criteria, such as land cover or elevation. Reclassification can be used to convert a continuous raster into a categorical raster or to simplify a complex raster.
- **Resampling**: This is the process of changing the resolution of the raster data by adjusting the size of the pixels. Resampling can be used to increase or decrease the resolution of the raster data. It is often necessary when merging two rasters with different resolutions or when transforming the raster to a new coordinate system.
- **Reprojection**: This is the process of changing the coordinate reference system of the raster data. Reprojection is often necessary when the raster data is in a different coordinate system than the other data layers used in a project. This operation can involve stretching, compressing, or rotating the raster data to align with the new coordinate system.
- **Clipping**: This is the process of extracting a subset of a raster based on a defined boundary. Clipping can be used to extract a region of interest from a larger raster or to mask out areas of the raster that are not relevant to the analysis. The result of clipping is a new raster that contains only the cells within the boundary.

Visualisation

Raster data visualization is an important aspect of working with geospatial data. It allows us to visually explore the data and identify patterns and relationships that might not be apparent from the raw data. There are several techniques that can be used to visualize raster data:

- **Color Maps**: Color maps are used to display raster data by assigning a different color to each pixel value. A color map can be used to highlight specific values or ranges of values in the data. For example, a color map might use shades of green to represent vegetation density or shades of blue to represent water depth.
- **Hillshading**: Hillshading is a technique used to create a 3D effect on a 2D map by simulating the shadows and highlights that would be cast by a light source. Hillshading is particularly useful for visualizing elevation data, as it allows the viewer to better understand the topography of the area.
- **Transparency**: Transparency is a technique used to overlay raster data on top of other data layers. By adjusting the transparency

of the raster layer, it is possible to see both layers at the same time, allowing the viewer to identify relationships and patterns between the layers.

- **Contour Lines**: Contour lines are a way of representing elevation data by drawing lines that connect points of equal elevation. Contour lines can be used to create a 2D representation of the 3D topography of an area.
- **Heat Maps**: Heat maps are a way of representing raster data by using colors to indicate the density or frequency of a particular phenomenon. For example, a heat map might be used to visualize the distribution of crime in a city or the concentration of air pollution.

Libraries

There are several Python libraries available for working with raster data, each with its own strengths and weaknesses. Here are a few of the most commonly used ones:

- **GDAL**: GDAL (Geospatial Data Abstraction Library) is a powerful open-source library for reading, writing, and manipulating raster data in a variety of formats. It provides a wide range of functions for processing and analyzing raster data, including reprojecting, resampling, and cropping. GDAL also supports reading and writing vector data, making it a versatile tool for working with both raster and vector geospatial data.
- **Rasterio**: Rasterio is a Python library built on top of GDAL that provides a more user-friendly and Pythonic interface for working with raster data. It offers a simplified way to read and write raster data, as well as tools for basic processing and analysis. Rasterio also supports raster data visualization using the Matplotlib library.
- **NumPy**: NumPy is a powerful numerical computing library that can be used for working with raster data. It provides a fast and efficient way to perform basic operations on raster data, such as mathematical calculations and array manipulations. NumPy can also be used in conjunction with other libraries like Matplotlib and SciPy to perform more advanced raster analysis.
- **OpenCV**: OpenCV (Open Source Computer Vision) is a library primarily used for image and video processing, but it can also be used for working with raster data. It offers a wide range of functions for image manipulation and analysis, including filtering,

edge detection, and feature extraction. OpenCV is particularly useful for analyzing satellite imagery, aerial photography, and other high-resolution raster data.

- **Xarray**: Xarray is a library built on top of NumPy and Pandas that provides a powerful way to work with multi-dimensional raster data. It allows users to label and manipulate raster data along multiple dimensions, making it easier to perform complex operations and analysis on large datasets. Xarray also provides built-in support for parallel processing, making it a good choice for working with large-scale raster data.

As we delve deeper into raster data using the rasterio library, we'll focus on how to read and plot raster images. This will lay the groundwork for understanding how raster data can be manipulated and visualized to extract valuable insights about our planet's surface.

Cattle Density

In this practical example, we'll be using the rasterio library, a key Python tool for raster data handling, to open and read a raster dataset. This particular dataset represents cattle distribution.

The first step is to import the rasterio library. Rasterio is built on top of GDAL and provides a more Pythonic way of working with raster data. It simplifies the process of reading, writing, and manipulating raster data.

`rasterio.open(...)` is used to open the raster file. In this case, it's a GeoTIFF file representing cattle distribution. The variable cattle is now a DatasetReader object, which holds the raster data and its metadata (like spatial dimensions, number of bands, coordinate reference system, etc.).

```
import rasterio

cattle = rasterio.open('cattle_rob.tif')
print(cattle.crs)
```

```
ESRI:54030
```

Rasterio, a Python library for raster processing, includes a show function that is specifically designed for the visualization of raster data. It is essentially a matplotlib imshow function built within a wrapper in Rasterio to account for Geospatial data. `show` takes a Rasterio dataset object (like the one returned by `rasterio.open`) and renders it into a plot. It interprets the raster values and translates them into a visual format, displaying different pixel values in different colors or shades.

The show method can display single-band (grayscale) or multi-band (color) raster data. For multi-band data, it can render the bands as RGB (Red, Green, Blue) images. Rasterio's `show` function is designed to be compatible with Matplotlib, a popular Python plotting library. This compatibility allows for further customization of the raster plot using Matplotlib's extensive plotting tools and features. When displaying the raster, show also considers the geographic metadata associated with the dataset. This means that the resulting plot can include spatial references, like axes with geographic coordinates, if the raster data includes such metadata.

```
from rasterio.plot import show

show(cattle)
```

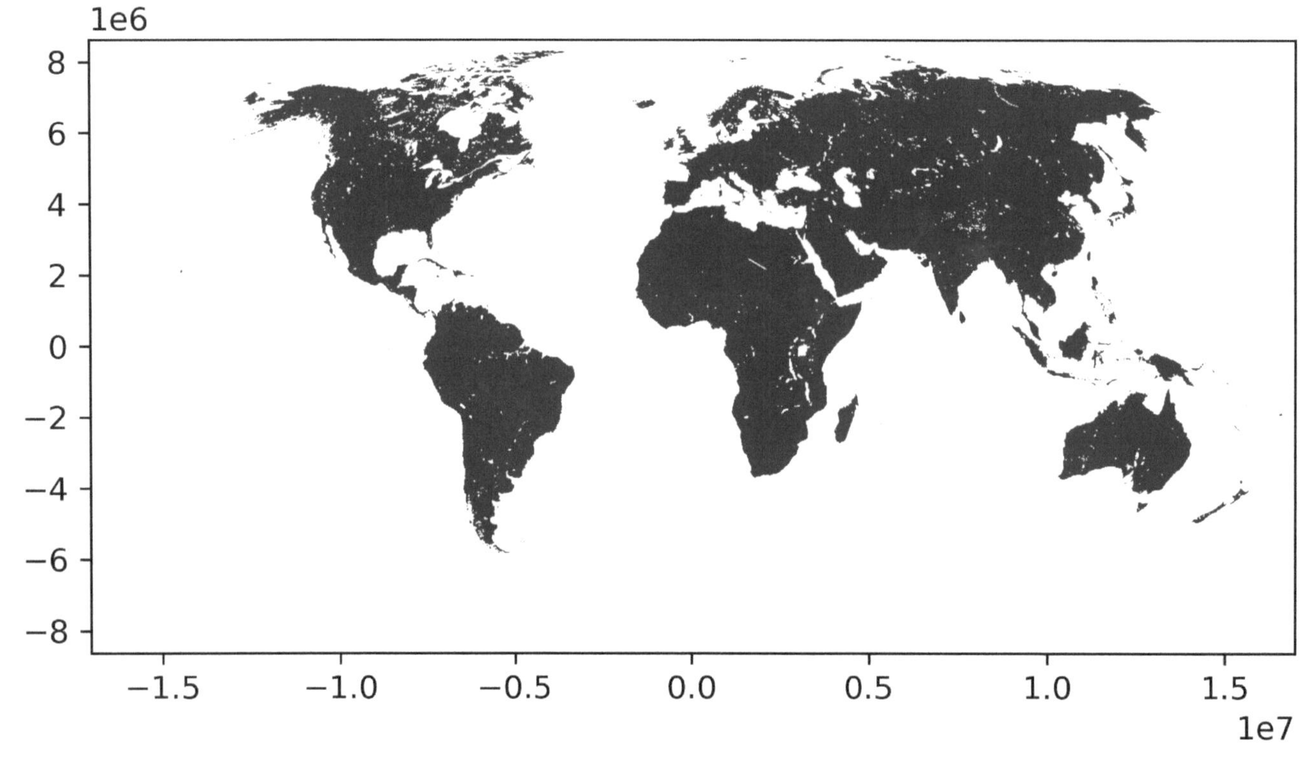

Figure 5.1: Output of raster.show for our cattle data without any customisations.

As in the previous example, we are going to load the world's country boundaries as a baselayer to use in the plot.

```
import geopandas as gpd

df = gpd.read_file('../PythonMapsData/NaturalEarth/10m_cultural/ne_10m_admin_0_countries.shp')
df = df[df['CONTINENT'] != 'Antarctica']
df['boundary'] = df.geometry.boundary
df['geometry'] = df['boundary'].copy()
```

Regardless of whether your data is geospatial, when plotting any sort of heatmap, it is often a good idea to construct a custom colourmap and colour boundaries for the plot. Colour boundaries are useful when your data is not linear and large or small values skew the visualisation. The code snippet provided demonstrates how to create a customized color map in Matplotlib. The bounds list defines the value ranges for which specific colors in the color map will be applied. Each number in bounds represents a boundary between color intervals. These were taken directly from those in the original research paper. (Gilbert, M., Nicolas, G., Cinardi, G. et al. Global distribution data for cattle, buffaloes, horses, sheep, goats, pigs, chickens and ducks in 2010. Sci Data 5, 180227 (2018).[10]

[10] https://doi.org/10.1038/sdata.2018.227)

reds is a base color map created using Matplotlib's built-in "Reds" color map. The number of colors in the color map is set to match the length of the bounds list. *newcolors* is an array of colors from the *Reds* color map. `np.linspace(0, 1, len(bounds))` generates evenly spaced numbers over a specified interval, which are used to fetch corresponding colors from the base color map.

`ListedColormap()` creates a new `ListedColormap` object from the array of newcolors. This custom color map is now ready to be applied to raster data. *norm* is a normalization object that maps the values in the raster data to the color map. BoundaryNorm uses the bounds list to determine how data values are mapped to the colors in the color map. The .N attribute of the color map (cmap.N) indicates the number of colors. In simple terms, the first colour in our colourmap will be used for all values between the first and second values in the bounds list.

```
import numpy as np
from matplotlib import cm
from matplotlib.colors import ListedColormap
from matplotlib import colors

bounds = [5, 10, 20, 50, 100, 250, 1000, 2500, 5000, 100000]

reds = cm.get_cmap("Reds", len(bounds))
newcolors = reds(np.linspace(0, 1, len(bounds)))
cmap = ListedColormap(newcolors)
norm = colors.BoundaryNorm(bounds, cmap.N)
```

We can then use Rasterio's show function to plot the cattle distribution raster data with our new custom color map (*cmap*) and normalization (*norm*). We are also going to overlay vector data (from a `GeoDataFrame` *df*) on top of the raster to add context to our data. The `transform=ccrs.Geodetic()` ensures that the vector data is correctly projected onto the Robinson projection, which is the base projection of the cattle raster. The combination of different data types in a single visualisation is going to be explored in depth in later chapters so don't worry that we are only touching on it here.

```
import matplotlib.pyplot as plt
from cartopy import crs as ccrs

fig = plt.figure(facecolor='white')
fig.set_size_inches(7, 3.5)
ax = plt.axes(projection=ccrs.Robinson())

show(cattle, ax=ax, cmap=cmap, norm=norm)
df.plot(ax=ax, edgecolor='black', lw=0.06, alpha=1, transform=ccrs.Geodetic())

ax.axis('off')
plt.show()
```

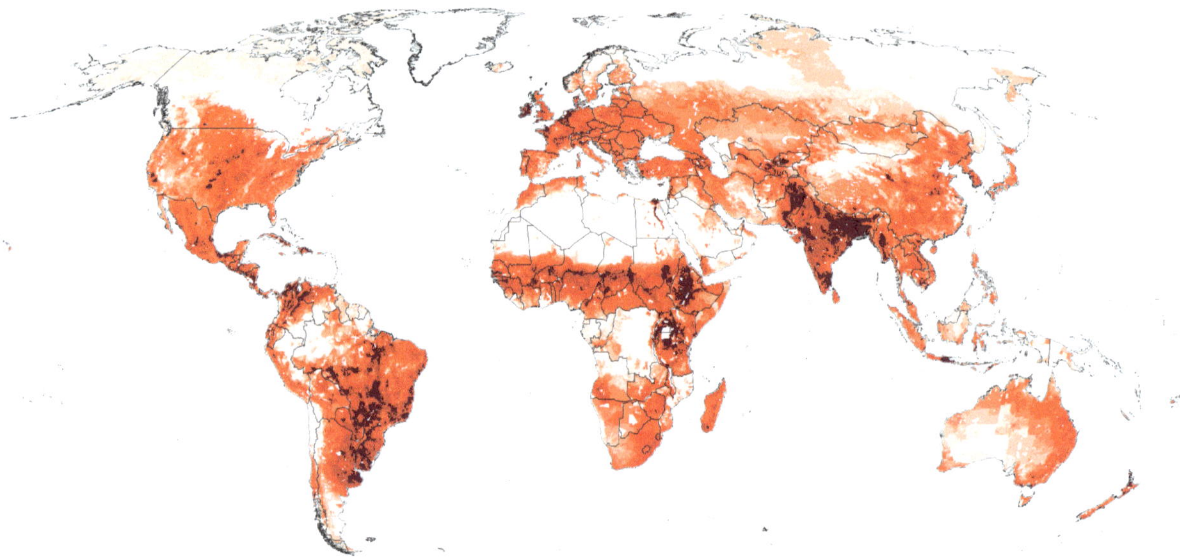

Figure 5.2: Cattle density of the world using our custom colourmap and colour boundaries.

Finally, we want to add a colourbar, so our audience knows what the colours mean. `cax` adds a new set of axes to the figure fig for the colorbar. The list *[0.5, 0.2, 0.15, 0.015]* defines the position and size of the colorbar axes as [left, bottom, width, height]. `cb_ticks` creates a list of tick positions for the colorbar. The ticks are calculated to be positioned at the midpoints between each pair of boundary values in the bounds array. `fig.colorbar(...)` creates the colorbar on the specified axes cax. `plt.cm.ScalarMappable(cmap=cmap, norm=norm)` maps the cmap and norm (normalization) specified earlier to the colorbar and `orientation="horizontal"` sets the colorbar orientation to horizontal. These lines define the colourbar, the following lines then customise it.

`cb.ax.set_xticklabels(...)` Ssets custom labels for the colorbar ticks, derived from the bounds list. `cb.outline.set_visible(False)` hides the outline of the colorbar for a cleaner appearance, `cb.ax.tick_params(...)` customizes the appearance of the ticks, including their size, color, and rotation, `plt.setp(...)` adjusts additional properties of the tick labels, such as their color and `cb.ax.set_xlabel(...)` Adds a label to the colorbar, in this case, 'Cattle Density', with custom font size, color, and padding.

```
import matplotlib.pyplot as plt
from cartopy import crs as ccrs

fig = plt.figure(facecolor='white')
fig.set_size_inches(7, 3.5)
ax = plt.axes(projection=ccrs.Robinson())

show(cattle, ax=ax, cmap=cmap, norm=norm)
df.plot(ax=ax, edgecolor='black', lw=0.06, alpha=1, transform=ccrs.Geodetic())

cax = fig.add_axes([0.5, 0.2, 0.15, 0.015])
cb_ticks = [bounds[i] + (bounds[i+1] - bounds[i]) / 2 for i in range(len(bounds) - 1)]
cb = fig.colorbar(plt.cm.ScalarMappable(cmap=cmap, norm=norm), cax=cax,
    orientation="horizontal", pad=0, format='%.3g', ticks=cb_ticks, drawedges=False)
cb.ax.set_xticklabels([str(val) for val in bounds[1:]])
cb.outline.set_visible(False)
cb.ax.tick_params(labelsize=4, width=0.2, length=0.75, rotation=60, color='black')
plt.setp(plt.getp(cb.ax, 'xticklabels'), color='black')
cb.ax.set_xlabel(f'Cattle Density', fontsize=5, color='black', labelpad=0)

ax.axis('off')
plt.show()
```

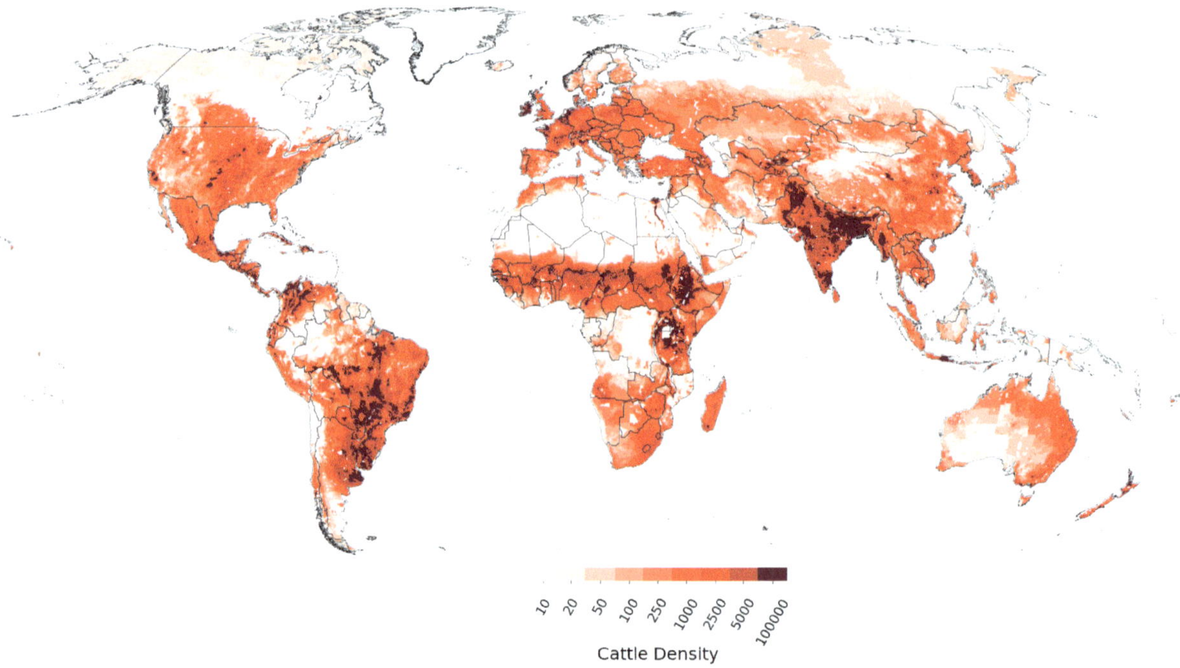

Figure 5.3: Cattle density of the world with a colourbar.

6

Mapping the World's Flight Paths

6.1 Converting Points to Lines - Airways

The previous chapters were all aimed at providing an introduction to the various spatial data types and formats with simple examples. This chapter marks a change of pace where we will start to apply our geospatial knowledge to generate more complicated maps. In this tutorial we are going to explore how to use the airports data that we encountered in chapter one and convert these into a map of the world's flight paths by combining them with a list of routes between airports. This is an exercise in data manipulation and data visualisation and you should come away with a better understanding of libraries like Pandas, GeoPandas, Matplotlib and Shapely.

The data for this exercise comes from the openflights database and is open source. The openflights database provides a list of airports in one file and a breakdown of the number of routes between each airport in another. So the dataset does not contain the specific coordinates of each route and that is what we will need to derive ourselves. In an ideal world we would have flight data consisting of position updates for every plane flying over the course of a year and we could build a 2D histogram of this data to show off flight paths. Unfortunately this data is difficult to come by so we have to make do with a list of airports and the number of routes between each. Our aim will be to create a `GeoDataFrame` of LineStrings to describe each route. The assumption that we make with this is that planes fly directly from airport to airport.

The first dataset is the airports data that we encountered in chapter one. As we have covered this already, we will not dwell too long on this. This data contains a list of airports and their locations as well as some other information. In total there are 7608 unique airports. The `read_csv` function from Pandas is employed to load this data, specifying a comma as the delimiter and defining custom column names like 'id', 'name', 'city', 'country', and others because this dataset does not have any column headers. It is important to note the airports iata, which is a unique identifier for each airport and will be used later to link airports in routes.

```
import pandas as pd

airports = pd.read_csv("airports",
                       delimiter=',',
                       names=['id', 'name', 'city', 'country', 'iata',
                              'icao', 'lat', 'long', 'altitude', 'timezone',
                              'dst', 'tz', 'type', 'source'])
```

The second dataset is a csv containing a list of routes. These routes take the form of iata pairs linking the source airport and the destination airport.

```
routes = pd.read_csv("routes",
                     delimiter=',',
                     names=['airline', 'id', 'source_airport', 'source_airport_id',
                            'destination_airport', 'destination_airport_id',
                            'codeshare', 'stops', 'equitment'])
```

We want to manipulate our data so that we have a DataFrame of Linestrings joining the source airport to the destination airport for each route. The method that we are going to employ to create LineStrings for each route is as follows.

- Create a duplicate airports `DataFrame` and label one as source and one as destination.
- Merge the source `DataFrame` with the routes using the routes source iata code.
- Merge the destination `DataFrame` with the routes using the routes destination iata code.
- Use the latitude and longitude values of the source and destinations airports which are now merged into the routes `DataFrame` to create a `LineString` representing that route.

Below we are duplicating the original airports `DataFrame` and labelling the columns of one with *_source* and the other with *_destination*.

```
source_airports = airports.copy()
destination_airports = airports.copy()
source_airports.columns = [str(col) + '_source' for col in source_airports.columns]
destination_airports.columns = [
    str(col) + '_destination'
    for col in destination_airports.columns
]
```

These two DataFrames are then merged with the routes `DataFrame` to give a `DataFrame` with the latitude/longitude values of the destination and source airports. Note the columns shown below, there are four columns: *lat_source, long_source, lat_destination* and *long_destination* for each route.

```
routes = routes[['source_airport', 'destination_airport']]
routes = pd.merge(routes, source_airports, left_on='source_airport',
                  right_on='iata_source')
routes = pd.merge(routes, destination_airports, left_on='destination_airport',
                  right_on='iata_destination')
```

```
Index(['source_airport', 'destination_airport', 'name_source', 'iata_source',
    'icao_source', 'lat_source', 'long_source', 'name_destination',
    'iata_destination', 'icao_destination', 'lat_destination',
    'long_destination', 'journey', 'count', 'geometry'],
    dtype='object')
```

The values are then used to generate a `GeoDataFrame` with LineStrings for each route using the `shapely.geometry.LineString` method. This takes a list of latitude and longitude pairs and creates a line that joins those points.

```
import geopandas as gpd
from shapely.geometry import LineString

geometry = [
    LineString([
        [routes.iloc[i]['long_source'], routes.iloc[i]['lat_source']],
        [routes.iloc[i]['long_destination'], routes.iloc[i]['lat_destination']]
    ]) for i in range(routes.shape[0])
]
routes = gpd.GeoDataFrame(routes, geometry=geometry, crs='EPSG:4326')
```

```
Index(['source_airport', 'destination_airport', 'name_source', 'iata_source',
    'icao_source', 'lat_source', 'long_source', 'name_destination',
    'iata_destination', 'icao_destination', 'lat_destination',
    'long_destination', 'journey', 'count', 'geometry'],
    dtype='object')
```

The data can now be plotted to give an airways map. Unfortunately it is not possible to simply plot the data as is because the curvature of the earth needs to be taken into account when plotting the lines. The map below shows direct lines between each airport which is incorrect.

```
fig = plt.figure()
ax = plt.axes()

routes.plot(ax=ax, color='black', linewidth=0.1)
ax.axis('off')
plt.show()
```

Cartopy can be used to manipulate the way that lines are plotted. The `transform=ccrs.Geodetic()` method transforms the LineStrings to account for the earths curvature. I have also changed the projection to the Robinson projection because it is one of the more realistic ways to

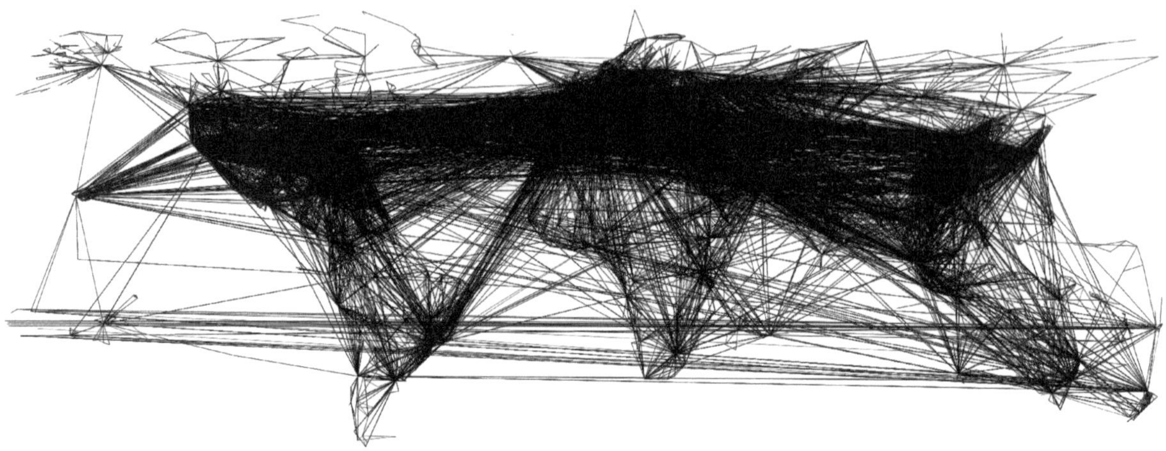

Figure 6.1: Map of the world's flight paths defined as the direct path between airports. The lines have been curved according to the earths surface.

plot global datasets, accounting for the fact that the earth is not flat. I also have reduced the alpha value to 0.1 to make it a little bit clearer.

```
fig = plt.figure(facecolor='black')

ax = plt.axes(projection=ccrs.Robinson())
ax.patch.set_facecolor('black')

routes.plot(ax=ax, transform=ccrs.Geodetic(), color='white', linewidth=0.1, alpha=0.1)

plt.setp(ax.spines.values(), color='black')
plt.setp([ax.get_xticklines(), ax.get_yticklines()], color='black')

plt.show()
```

The final step for this section is to plot these airways on top of the worlds airports. We are going to add a twist though and scale the size of the airports to be based on the number of flights that land there. We are currently applying a single alpha value to the plot, but another way to visualise the data is to provide a customised value depending on how frequent the route is used.

To do this we need to do some more Pandas manipulations. First we are going to create a unique identifier for each route by adding a new column to the 'routes' `DataFrame` that concatenates the source and destination IATA codes, creating a unique identifier for each route. Then we employ NumPy's `unique` function to find all unique journeys and their occurrence counts and convert the unique routes and their counts into a new `DataFrame`. Finally, we combine this new `DataFrame` with the original 'routes' `DataFrame` to associate each route with its frequency.

```
import numpy as np

```

Figure 6.2: Map of the world's airports scaled according to the number of flights that land there and the world's airways.

```
routes['journey'] =  routes['iata_source'] + '_' + routes['iata_destination']
journeys, counts = np.unique(routes['journey'], return_counts=True)
counts = pd.DataFrame({'journey': journeys, 'count': counts})
routes = pd.merge(routes, counts, on='journey')

routes['alpha'] = 1 / (routes['count'] * 3)

single_routes = routes.drop_duplicates("journey")
```

Now we need to analyze and aggregate data regarding the usage frequencies of airports as source and destination points in a flight routes dataset. First we calculate the number of times each airport appears as a source and as a destination in the routes dataset, using the `value_counts()` method from Pandas. These counts are then transformed into separate DataFrames, *source_counts* and *destination_counts*, and are reset into a structured format with appropriate column names. Next, these two DataFrames are merged into a single `DataFrame`, *airport_counts*, based on their common indices, which presumably represent airport identifiers. The code then adds a new column to this merged `DataFrame`, summing up the source and destination counts to give a total usage count for each airport. In the final step, the *airport_counts* `DataFrame` is merged with another `DataFrame` named airports, which likely contains more detailed information about each airport. This merge enriches the frequency data with additional details, providing a comprehensive view of how frequently each airport is used in the network, both as a source and a destination.

We then create a Geopandas `DataFrame` from the data and create a marker size column which is based upon about 1/25 the count.

```
# Calculate counts for source and destination airports
source_counts = (routes.source_airport.value_counts().
                 reset_index(name='source_airport_count'))
destination_counts = (routes.destination_airport.value_counts().
                      reset_index(name='destination_airport_count'))

# Merge the source and destination counts
airport_counts = pd.merge(source_counts, destination_counts, left_on='index',
                          right_on='index')

# Calculate total counts
airport_counts['count'] = (airport_counts['source_airport_count'] +
                           airport_counts['destination_airport_count'])

# Merge with airports dataframe
airport_counts = pd.merge(airport_counts, airports, left_on='index', right_on='iata')
```

We then create a Geopandas `DataFrame` from the data and create a marker size column which is based upon about 1/25 the count.

```
from shapely.geometry import Point

geometry = [Point(xy) for xy in zip(airport_counts.long,
                                    airport_counts.lat)]
airport_counts = gpd.GeoDataFrame(airport_counts,
                                  geometry=geometry,
                                  crs="EPSG:4326")

airport_counts['markersize'] = airport_counts['count'] / 25
```

All we need to do now is put then both together and plot our routes dataframe and our airports `DataFrame`.

```


fig, ax = plt.subplots(subplot_kw={'projection': ccrs.Robinson()},
                       figsize=(6,9), facecolor="black")
ax.patch.set_facecolor('black')
single_routes.plot(ax=ax,
                   transform=ccrs.Geodetic(),
                   color='white',
                   linewidth=single_routes['alpha'],
                   alpha=0.2)
airport_counts.plot(ax=ax,
                   transform=ccrs.PlateCarree(),
                   markersize=airport_counts['markersize'],
                   alpha=0.8,
                   column=airport_counts['long'],
                   cmap='jet',
                   edgecolors='none')

plt.setp(ax.spines.values(), color='#090909')
plt.setp([ax.get_xticklines(), ax.get_yticklines()], color='#090909')

plt.show()
```

Figure 6.3: Map of the world's airports scaled according to the number of flights that land there and the world's airways.

7

Mapping the World's Emissions

7.1 Advanced customisations - Plotting Global Emissions

In this chapter, we delve into the complex and pressing topic of global CO_2 emissions, a subject that sits at the heart of contemporary environmental concerns. Our objective is not just to map these emissions, but to do so in a way that reveals deeper insights and patterns that might otherwise be obscured in more conventional representations. This chapter is dedicated to showcasing advanced techniques in data visualization, particularly focusing on the nuanced portrayal of CO_2 emissions across the Earth's grid squares.

In this section, we commence our analysis by importing the global CO_2 emissions data using the pandas library's read_csv function. This dataset is structured into three key columns: 'lat' for latitude, 'lon' for longitude, and 'emission', which represents the CO_2 emission values. A quick printout of this data gives us a preliminary view of its structure, confirming that we have the necessary information for our mapping exercise.

```
import pandas as pd

co2 = pd.read_csv("../../HumanImpact/emissions/data/vco2.txt", delimiter=';')

print(co2)
```

```
           lat    lon    emission
0         88.1  -50.7     11.0301
1         88.1  -50.6     15.7066
2         88.1  -50.5     19.1943
3         88.1  -50.4     22.0621
4         88.1  -50.3     24.5536
...        ...    ...         ...
2191880  -77.8  166.5  11459.2000
2191881  -77.8  166.6    694.4980
2191882  -77.9  166.3   2133.1000
```

```
2191883 -77.9  166.4    1389.0000
2191884 -77.9  166.5    7887.5100

[2191885 rows x 3 columns]
```

Next, we turn to an initial visualization of this data. We employ a basic scatter plot to create a straightforward, point-by-point representation of the emissions on a global scale. This scatter plot is generated using standard Python plotting libraries and serves as a rudimentary visual interpretation of the data. Each point on the plot corresponds to a grid square's location on the globe, with its color and/or size indicating the emission level at that point.

This initial plotting step is crucial as it provides us with a baseline visualization. However, this basic scatter plot may not adequately convey the complexities and scale differences in the emission data, especially when dealing with a wide range of values.

```
import matplotlib.pyplot as plt

fig = plt.figure()
fig.set_size_inches(7, 3.5)
plt.scatter(co2['lon'], co2['lat'], s=0.05,  edgecolors='none')

plt.show()
```

When visualizing data like global CO_2 emissions, where values can vary over several orders of magnitude, a linear scale may not always be the most effective way to represent the data. This is particularly true when the dataset contains both very high and very low values. In a linear representation, extreme values at the higher end can overshadow subtler variations at the lower end, making it difficult to discern important patterns across the majority of the data range.

A logarithmic scale, or log scale, addresses this issue by transforming the scale of the data in such a way that the vast range of values is compressed into a more manageable and visually discernible scale. This transformation is particularly useful for:

- Log scales enhance the visibility of smaller values, which would be less noticeable or even invisible on a linear scale.
- It prevents high-value data points from dominating the visual narrative, ensuring a more balanced representation across the entire range of values.
- By using a log scale, one can more easily spot trends and patterns that would be obscured in a linear scale, especially in datasets with exponential growth or decay characteristics.

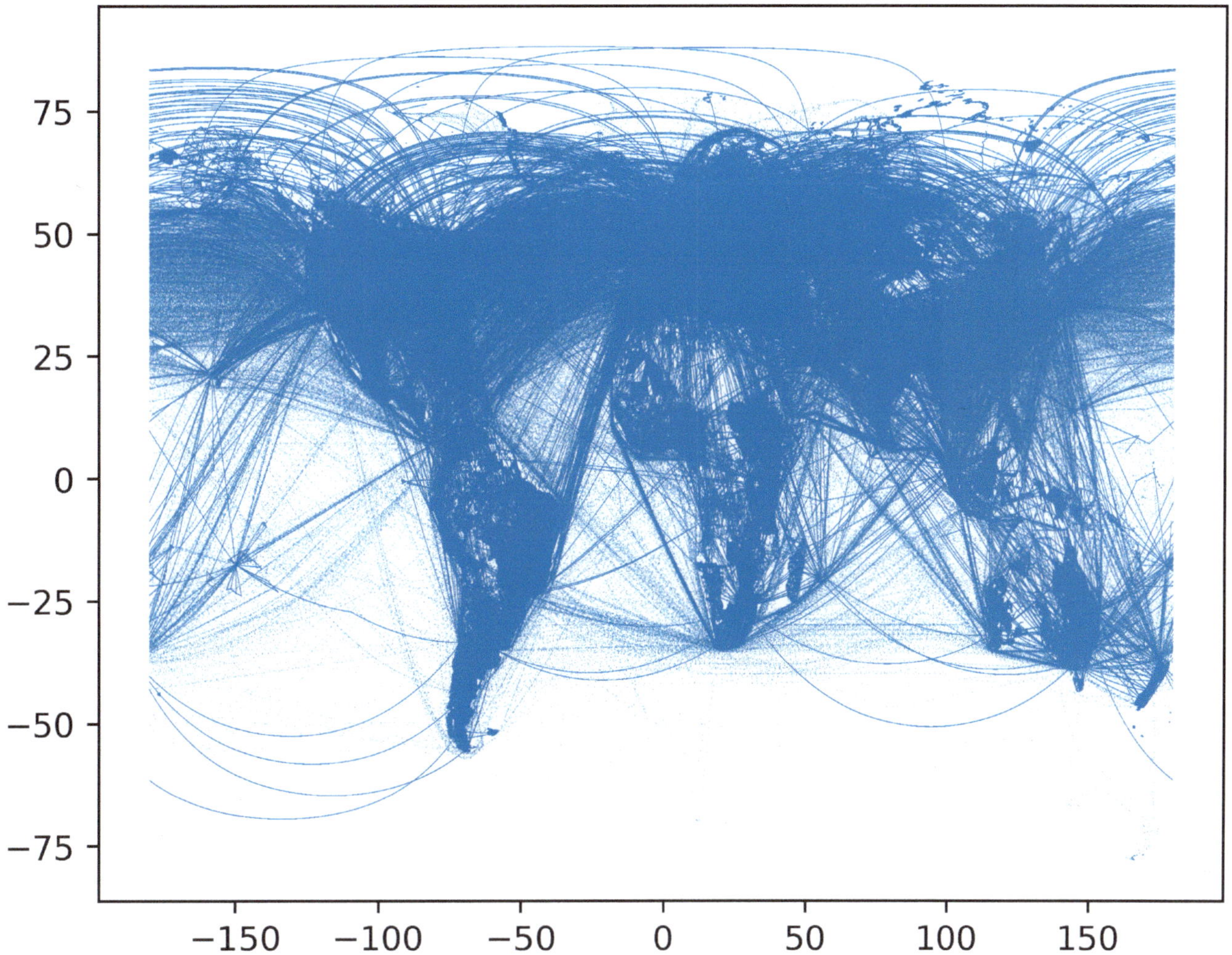

Figure 7.1: Map of the world's carbon dioxide emissions.

`matplotlib.colors.LogNorm()` is used to implement a logarithmic scale for color representations in plots. `LogNorm()` normalizes the input data on a logarithmic scale. This means that instead of mapping the raw emission values directly to colors, it maps their logarithms. This approach is particularly effective for data that spans several orders of magnitude, as in the case of CO_2 emissions. When used in conjunction with color maps (colormaps) in Matplotlib, `LogNorm()` ensures that the colors assigned to different data points represent the logarithmic value of the data, rather than the raw data. This results in a plot where the color variation more accurately reflects the relative differences in the logarithmic scale of the data. By employing `LogNorm()`, you can create heatmaps, scatter plots, or any other plot type where color intensity represents data magnitude, in a way that brings out nuances and patterns across the entire range of your dataset, not just the high or low extremes.

```
from matplotlib import colors

fig = plt.figure()
plt.scatter(co2['lon'], co2['lat'], s=0.05, edgecolors='none', c=co2['emission'],
            norm=colors.LogNorm(), cmap='hot_r')
plt.show()
```

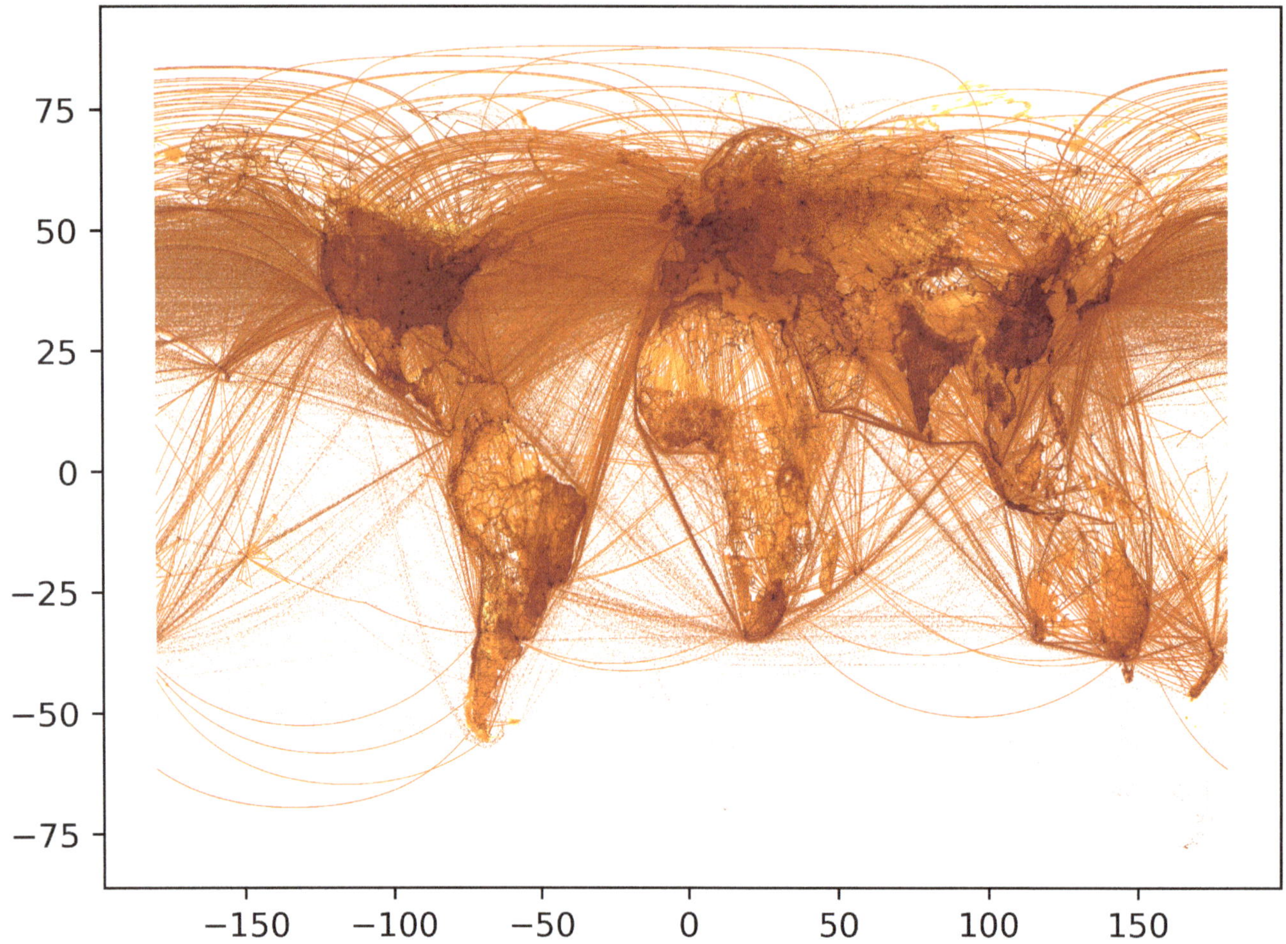

Figure 7.2: Map of the world's carbon dioxide emissions coloured according the emission value.

In the next phase we revisit the approach from chapter one of converting our latitude and longitude values into a `GeoDataFrame` of Points. We transform the latitude and longitude data into Shapely `Point` objects, enabling their use in spatial operations. These points are then integrated into a `GeoDataFrame`, which combines our emissions data with its geographic coordinates, facilitating geospatial analysis. Utilizing Cartopy, we reproject the data into the Robinson projection, a choice that balances the global representation. This step is crucial for accurate and meaningful visualization on a world map. With the data properly formatted and reprojected, we proceed to plot the map, showcasing the geographic distribution of CO_2 emissions in a visually effective manner (figure 7.3).

```
import geopandas as gpd
from shapely.geometry import Point

geometry = [Point(xy) for xy in zip(co2['lon'], co2['lat'])]
geodata = gpd.GeoDataFrame(co2, crs="EPSG:4326", geometry=geometry)
```

```
from cartopy import crs as ccrs

fig, ax = plt.subplots(facecolor='black', subplot_kw={'projection': ccrs.Robinson()})
ax.patch.set_facecolor('black')
fig.set_figwidth(6)
ax = geodata.plot(ax=ax, column='emission', transform=ccrs.PlateCarree(),
                  cmap='cividis', norm=colors.LogNorm(), s=0.05, edgecolors='none')
plt.show()
```

Figure 7.3: Map of the world's carbon dioxide emissions coloured according the emission value.

In the original research, the authors who produced a map with custom colour boundaries. We are going to employ the same colour boundaries and use the method outlined in chapter four to create a custom colourmap and colour boundaries. *bounds* is a list defining the boundaries for the colormap. These values represent the CO_2 emission levels at which the color changes. We then create a custom colourmap with 11 colours. `cm.get_cmap('afmhot', len(bounds))` retrieves the *afmhot* colormap with 11 discrete colors. The 'afmhot' colormap is known for its good perceptual uniformity and readability by those with color vision deficiency. `np.linspace(0, 1, len(bounds))` generates 11 evenly spaced numbers over the interval [0, 1], which are then used to extract specific colors from the 'cividis' colormap to create newcolors. The line `newcolors = newcolors[1:]` adjusts the colormap by removing the first color. This modification is done to better match the color scheme of the original dataset's map.

The `colors.BoundaryNorm(bounds, our_cmap.N)` creates a normalization object (norm) that maps the data points to the discrete colormap based on these defined boundaries.

With the custom colormap (*our_cmap*) and the normalization object (*norm*) defined, you replot your map. This replotting applies the customized color scheme to your CO_2 emissions data, ensuring that the map visually corresponds to the distinct color boundaries of the original dataset.

```
import numpy as np
from matplotlib import cm
from matplotlib.colors import ListedColormap

bounds = [0.0, 0.06, 6, 60, 600, 3000, 6000, 24000, 45000, 120000]
our_cmap = cm.get_cmap('afmhot', len(bounds))
newcolors = our_cmap(np.linspace(0, 1, ln(bounds)))
newcolors = newcolors[1:]
norm = colors.BoundaryNorm(bounds, our_cmap.N)
```

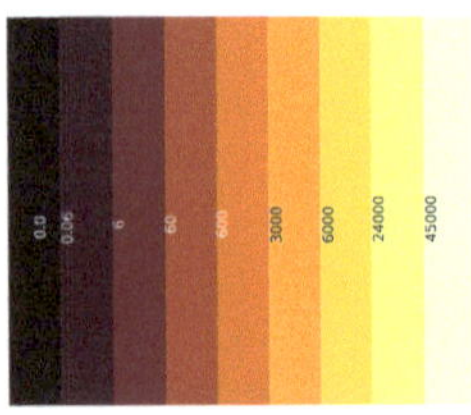

Figure 7.4: Illustration of how colourmaps and colour boundaries work.

```
fig, ax = plt.subplots(facecolor='black', subplot_kw={'projection': ccrs.Robinson()})
ax.patch.set_facecolor('black')

ax = geodata.plot(ax=ax, column='emission', transform=ccrs.PlateCarree(),
                  cmap=our_cmap, norm=norm, s=0.05, alpha=1, edgecolors='none')

plt.show()
```

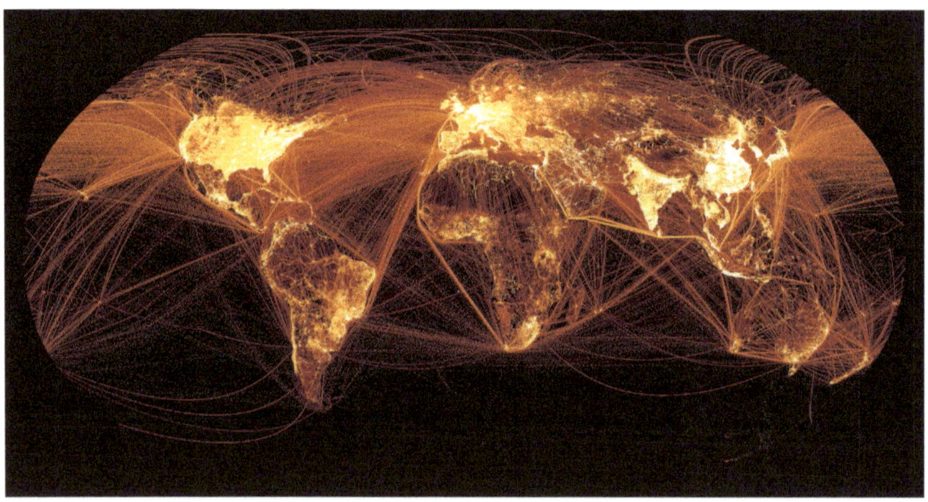

Figure 7.5: Map of the world's carbon dioxide emissions coloured according the emission value.

To add a colorbar, a new axes (cax) is created on the figure to hold the colorbar. `plt.cm.ScalarMappable` creates a scalar mappable object (*sm*) with the same colormap and normalization as used in the plot.

`fig.colorbar(...)` adds the colorbar to cax. The colorbar's appearance is customized with its orientation, padding, tick format, and edge visibility. Custom ticks are defined to represent specific emission levels, and the colorbar's tick labels are styled and sized.

The colorbar's label is set to display: CO_2 Tons/Year, indicating the metric used for the emissions data. See the final map, figure 7.6.

```
fig, ax = plt.subplots(facecolor='black', subplot_kw={'projection': ccrs.Robinson()})
ax.patch.set_facecolor('black')

ax = geodata.plot(ax=ax, column='emission', transform=ccrs.PlateCarree(),
                  cmap=our_cmap, norm=norm, s=0.05, alpha=1, edgecolors='none')

plt.show()
```

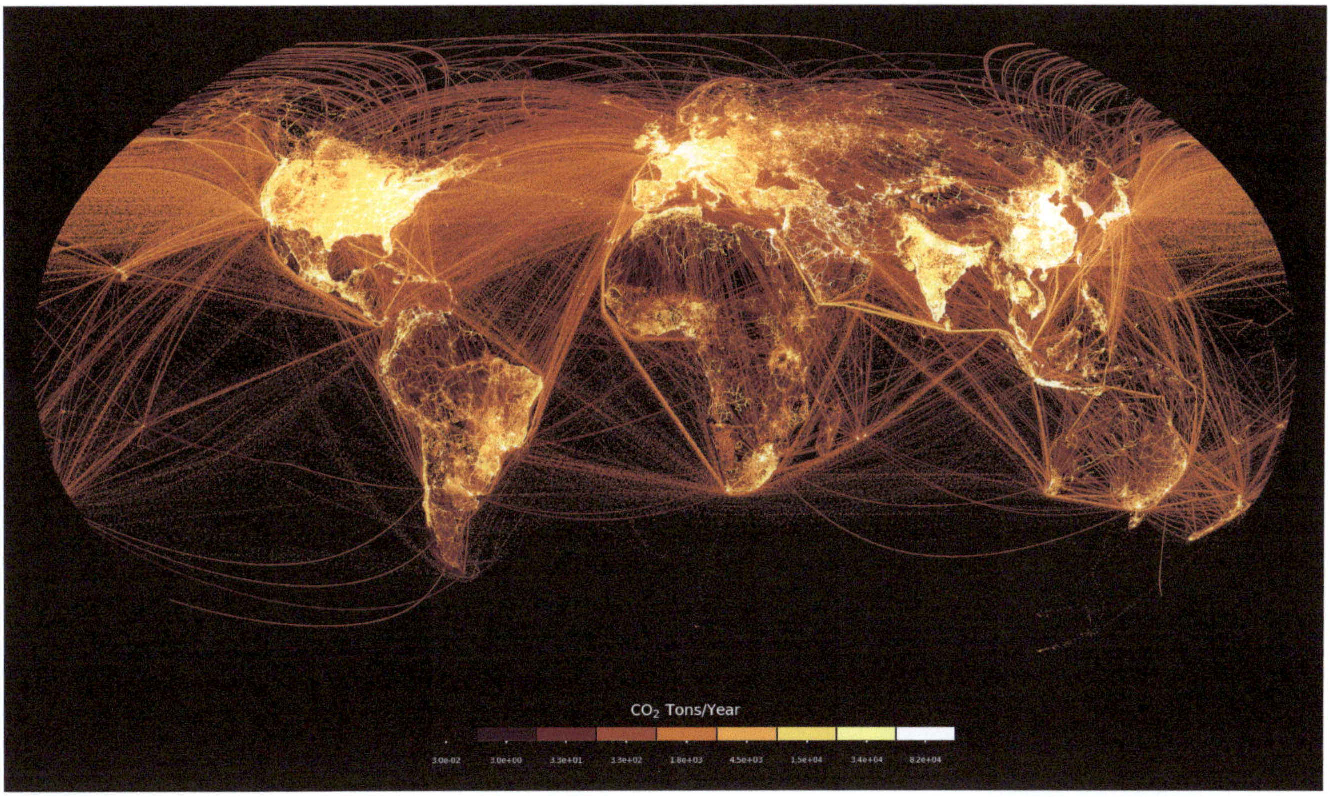

Figure 7.6: Final map of the world's carbon dioxide emissions coloured according the emission value.

8

Mapping the World's Tropical Storms

8.1 Plotting Tropical Storms

As we have journeyed through the earlier chapters of this book, the Robinson projection has been our constant companion, offering a balanced and aesthetically pleasing view of the Earth. This projection, with its unique approach to balancing size and shape distortions, has served as a practical and visually appealing backdrop for our explorations. However, the world of cartography is rich with diverse map projections, each bringing its own perspective and insights to the representation of our planet.

In this chapter, we delve deeper into the realm of map projections, moving beyond the familiar confines of the Robinson projection. Our focus shifts to three fundamental types of projections: planar (or azimuthal), cylindrical, and conical. This exploration is not just a theoretical exercise; it is a practical journey, as we will apply these projections to the visualization of historical tropical storms. These storms, intriguing in their patterns and significant in their impact, do not cross the equator, presenting a unique opportunity to examine the strengths and limitations of different projection types.

Building on the foundation laid in previous chapters, we will first provide a detailed overview of each projection type, discussing its key characteristics and the specific aspects of Earth's geometry it best represents. Accompanied by Python code snippets using the Cartopy library, we will then illustrate how to plot historical tropical storms using each type of projection. This hands-on approach will not only enhance your understanding of the theoretical aspects of map projections but also demonstrate their practical applications in visualizing complex meteorological data.

As we explore these various projections, remember that each map projection is a tool with specific uses and limitations. The choice of a projection often involves balancing accuracy in area, shape, distance, and direction with the intended use of the map. Through the lens of tropical storm tracking, we will see how these projections influence our perception and interpretation of geographic information, reminding us of the importance of selecting the right projection for the task at hand.

This chapter aims to expand your cartographic toolkit, equipping you with the knowledge and skills to make informed choices about map projections in your future endeavors. Let us embark on this enlightening journey, enhancing our understanding of how we map our world and the powerful stories these maps can tell.

As in previous examples we load the natural earth countries shapefile as our background image.

```
import geopandas as gpd

countries = gpd.read_file('NaturalEarth/10m_cultural/ne_10m_admin_0_countries.shp')
```

In this section, we start by loading and preparing the historical tropical storm data, which forms the basis for our map projections. The data is distributed across multiple CSV files. We use glob to generate a list of these files. Once the files are identified, we utilize pandas to read and concatenate these files into a single `DataFrame`. This consolidation is crucial for ease of analysis and visualization, as it brings all the storm data into one structured format. During this process, specific column names are provided to maintain consistency and clarity in our dataset.

```
import glob
import pandas as pd

files = glob.glob("STORM_data/*.txt")

cols = ["Year", "Month", "TC number", "Time step", "Basin", "Latitude", "Longitude",
        "Minimum pressure", "Maximum wind speed", "Radius to maximum winds", "Category",
        "Landfall", "Distance to land"]

data = pd.concat(pd.read_csv(f, delimiter=",", header=None, names=cols) for f in files)
```

The next step we filter out any data points with invalid latitude or longitude values, as these would lead to inaccuracies in our mapping. Additionally, we apply a temporal filter to focus on more recent events, specifically limiting our analysis to storms that occurred within the last 100 years. This choice is made to keep the data relevant to current geographical and climatic contexts.

```
hundred = data.loc[data['Year'] > 949]
dodgy = hundred.loc[hundred['Longitude'] > 180]
dodgy['Longitude'] = dodgy['Longitude']  - 360
hundred = hundred.loc[hundred['Longitude'] < 180]

hundred = pd.concat([hundred, dodgy])

hundred = hundred.loc[hundred['Category'] > 1]
```

Finally, we transform the latitude and longitude data into geometric points. With the data now in a clean, structured, and geographically aware format, we can proceed to illustrate the intricacies and impacts of different map projections using these historical tropical storm tracks as our case study.

```
from shapely.geometry import Point

geometry = [Point(xy) for xy in zip(hundred['Longitude'], hundred['Latitude'])]
geodata = gpd.GeoDataFrame(hundred, crs="EPSG:4326", geometry=geometry)
```

Having established a solid foundation in handling and preparing our historical tropical storm data, we now embark on the first step of our cartographic exploration: visualizing this data using the Robinson projection. As you might recall from previous chapters, the Robinson projection is a compromise projection, skillfully balancing the representation of area, shape, distance, and direction. This makes it an ideal starting point for visual comparisons with other map projections.

The Robinson projection's balanced view provides a familiar and aesthetically pleasing global perspective, one that we have utilized in various examples so far. By plotting our tropical storm data on this projection, we can observe the global spread and trajectories of these storms in a context that is both geographically coherent and visually intuitive. This visualization will serve as a baseline – a point of reference against which we can compare other types of projections.

Once we have established this baseline, we will dive into the heart of this chapter: exploring different types of map projections. Our journey will take us through three primary categories – planar (or azimuthal), cylindrical, and conical projections. Each of these categories offers unique perspectives and varying degrees of distortion in aspects like scale, area, and angular relationships.

For each type of projection, we will provide a detailed overview of its characteristics, followed by practical examples using the Cartopy library in Python. These examples will not only illustrate the theoretical principles of each projection but also demonstrate their practical applications in mapping the paths and impacts of historical tropical storms.

By juxtaposing these different projections with our initial Robinson projection map, we will gain a deeper understanding of how the choice of projection can influence our perception and interpretation of geographical data. This comparative approach will highlight the importance of selecting the most appropriate projection for the specific requirements of a mapping task, particularly in the context of complex meteorological phenomena like tropical storms.

Let's begin by revisiting the Robinson projection with our tropical storm data, setting the stage for a comprehensive exploration of the diverse world of map projections.

```
import matplotlib.pyplot as plt
import cartopy.crs as ccrs

fig, ax = plt.subplots( subplot_kw={'projection': ccrs.Robinson()})
fig.set_size_inches(10, 10)
ax.axis('off')
countries.plot(ax=ax, color="black", transform=ccrs.PlateCarree())
geodata.plot(ax=ax, markersize=1, alpha=1, edgecolors='none',
             column="Maximum wind speed", cmap="jet", transform=ccrs.PlateCarree())
ax.gridlines(crs=ccrs.PlateCarree(), draw_labels=True,
                         linewidth=1, color='gray', alpha=0.6, linestyle='--')
plt.show()
```

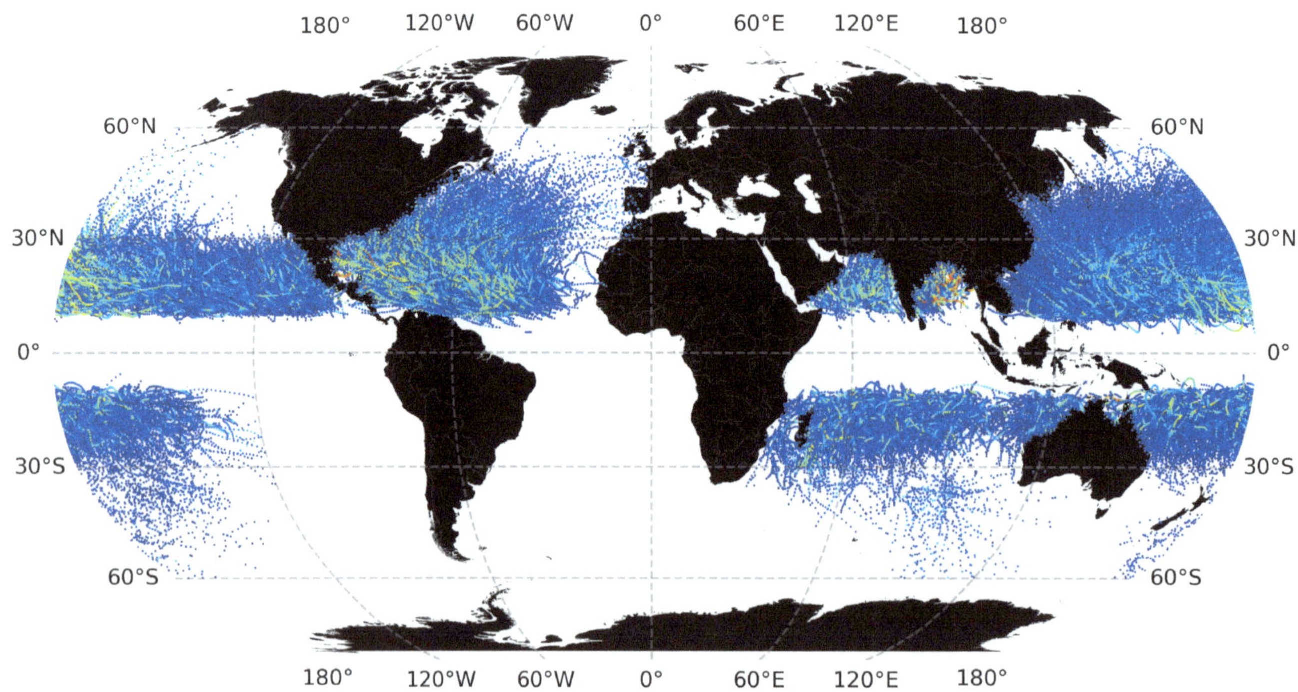

Figure 8.1: Tropical storms, plotted in the Robinson projection.

Planar projections

A planar projection, also known as an azimuthal projection, is a type of map projection that projects the Earth's surface onto a flat plane tangential to a point or a plane. This means that the center of the projection represents the point of tangency, and all other points are projected outward from that point. The resulting map is often circular in shape, with distortions increasing as the distance from the center of the projection increases.

Planar projections are commonly used for depicting the poles, since they preserve the direction from the center point to all other points on the map. However, they are not well-suited for representing large areas because the distortions increase rapidly as you move away from the center point. Planar projections can also be used to represent other regions of the Earth, such as equatorial regions, but they are generally less commonly used for this purpose.

There are several different types of planar projections, including the following:

- **Gnomonic projection**: This is the most basic planar projection, which projects the Earth's surface onto a plane that is tangent to a single point on the Earth's surface. This projection preserves directions, but distorts distances and areas.
- **Stereographic projection**: This projection projects the Earth's surface onto a plane that is tangent to a single point on the Earth's surface. It preserves angles and shapes, but distorts distances and areas.
- **Orthographic projection**: This projection projects the Earth's surface onto a plane that is tangent to the Earth's surface at a single point. It preserves shapes, but distorts distances and areas.
- **Azimuthal equidistant projection**: This projection projects the Earth's surface onto a plane that is tangent to the Earth's surface at a single point. It preserves distances and directions from the center point, but distorts other properties such as area and shape.

In summary, planar projections are a type of map projection that projects the Earth's surface onto a flat plane tangential to a point or a plane. They are commonly used for depicting the poles, but are less well-suited for representing large areas due to their rapid distortions as you move away from the center point. There are several different types of planar projections, each with its own advantages and disadvantages depending on the purpose of the map.

```
fig, ax = plt.subplots(subplot_kw={'projection': ccrs.AzimuthalEquidistant()})
fig.set_size_inches(10, 10)
countries.plot(ax=ax, color="black", transform=ccrs.PlateCarree())
geodata.plot(ax=ax, markersize=1, alpha=1, edgecolors='none',
             column="Maximum wind speed", cmap="jet",
             transform=ccrs.PlateCarree())
ax.gridlines(crs=ccrs.PlateCarree(), draw_labels=True,
                      linewidth=1, color='gray', alpha=0.8, linestyle='--')
plt.show()
```

Conical Projection

A conical projection is a type of map projection that projects the Earth's surface onto a cone. The cone is then unrolled to produce a flat map. The resulting map typically has curved horizontal lines that represent lines of latitude, and straight vertical lines that represent lines of longitude.

Conical projections are less common than cylindrical projections, but they are still used today, particularly for mapping mid-latitude regions that are long in the east-west direction. They are also used for mapping small areas such as countries or regions, because they produce less distortion than other types of projections.

Conical projections are generally classified as either tangent or secant, depending on whether the cone intersects the Earth's surface at one point or two points. A tangent conical projection has one point of tangency with the Earth's surface, while a secant conical projection has two points of tangency.

There are several different types of conical projections, including the following:

- **Lambert conformal conic projection**: This is a common conical projection that is often used for mapping mid-latitude regions, such as the United States. It is a conformal projection, which means that it preserves shapes and angles well, but distorts sizes and distances.
- **Albers equal-area conic projection**: This is a conical projection that preserves areas well, but distorts shapes and distances. It is often used for mapping regions with irregular shapes, such as states or provinces.
- **Polyconic projection**: This is a conical projection that uses a series of cones to project the Earth's surface onto a flat map. It is often used for mapping mountainous regions or small areas, be-

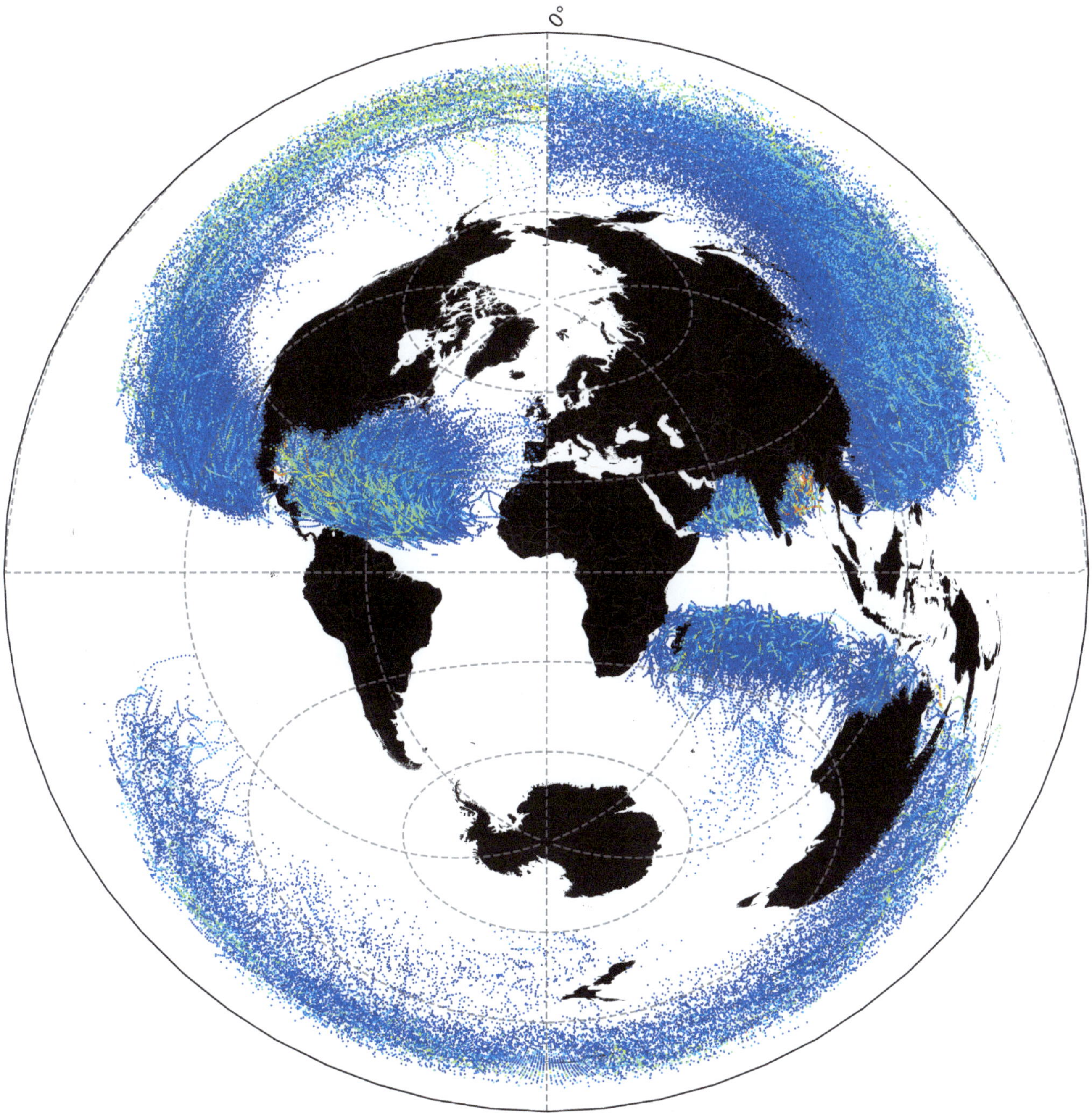

Figure 8.2: Tropical storms, plotted in the Azimuthal equidistant projection.

cause it produces less distortion than other types of projections.

In summary, conical projections are a type of map projection that projects the Earth's surface onto a cone, which is then unrolled to produce a flat map. They are less common than cylindrical projections, but they are still used for mapping mid-latitude regions and small areas, because they produce less distortion than other types of projections. There are several different types of conical projections, including the Lambert conformal conic projection, the Albers equal-area conic projection, and the polyconic projection.

```
fig, ax = plt.subplots(subplot_kw={'projection': ccrs.LambertConformal()})
fig.set_size_inches(10, 10)

countries.plot(ax=ax, color="black", transform=ccrs.PlateCarree())
geodata.plot(ax=ax, markersize=1, alpha=1, edgecolors='none',
             column="Maximum wind speed", cmap="jet",
             transform=ccrs.PlateCarree())
ax.gridlines(crs=ccrs.PlateCarree(), draw_labels=True,
                     linewidth=1, color='gray', alpha=0.8, linestyle='--')
plt.show()
```

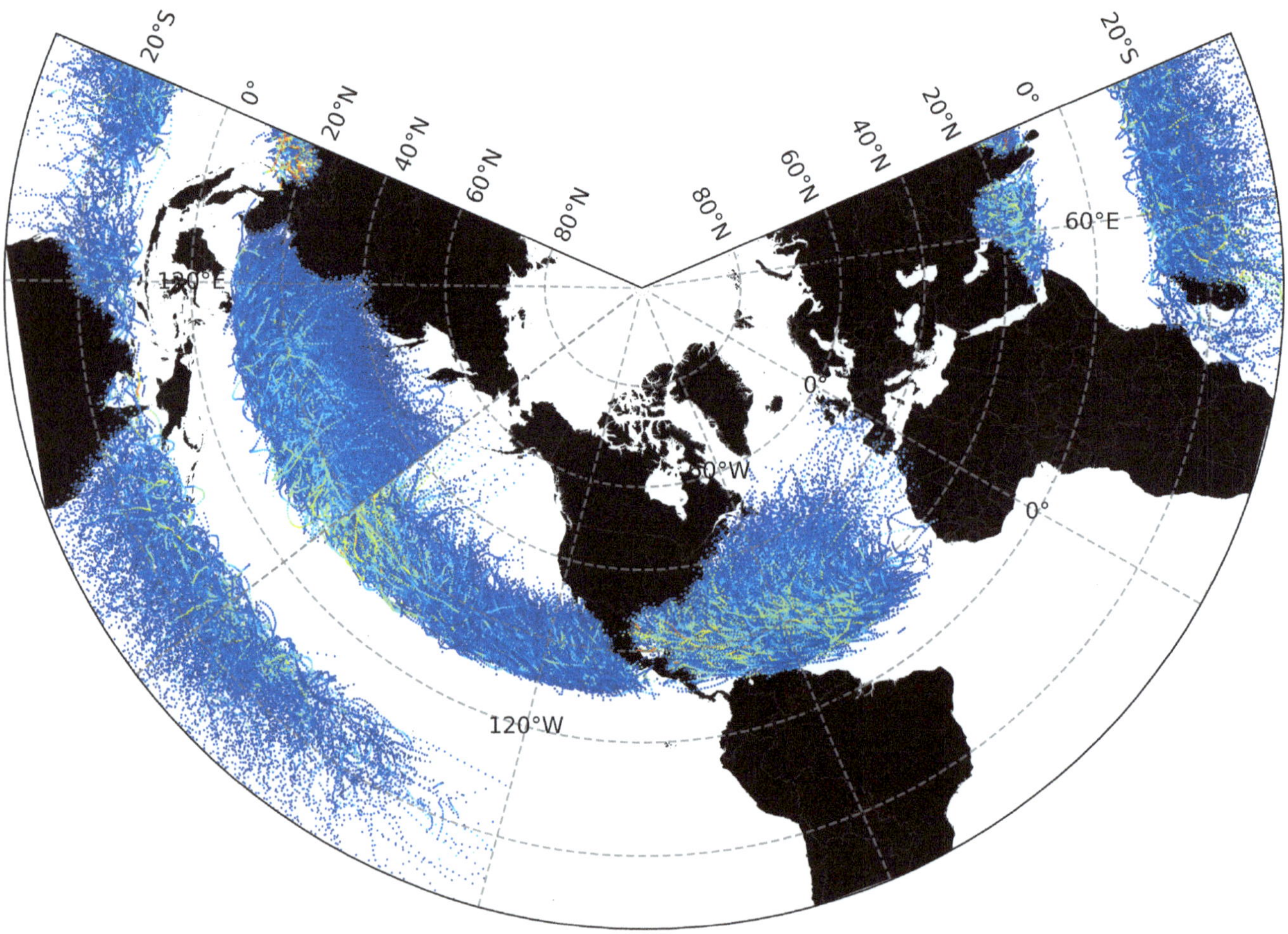

Figure 8.3: Tropical storms, plotted in the Lambert conformal conic projection.

Cylindrical Projection

A cylindrical projection is a type of map projection that projects the Earth's surface onto a cylinder. The cylinder is then unrolled to produce a flat map. The resulting map typically has straight horizontal lines that represent lines of latitude, and straight vertical lines that represent lines of longitude.

Cylindrical projections are one of the oldest types of map projections, and were used extensively by ancient Greek and Roman cartographers. They are still widely used today, particularly for world maps and maps of large regions, because they preserve angles and shapes well and are relatively simple to produce and use.

However, cylindrical projections also have significant distortions in both distance and area, particularly at the poles, where the distortion is severe. This is because the Earth's surface is a sphere, while a cylinder is a flat surface, so some distortion is inevitable.

There are several different types of cylindrical projections, including the following:

- **Mercator projection**: This is the most famous cylindrical projection, which was developed by Flemish cartographer Gerardus Mercator in 1569. It is a conformal projection, which means that it preserves shapes and angles, but greatly distorts sizes and distances, particularly near the poles. The Mercator projection is widely used for navigation and in schools because it makes it easy to draw straight lines on the map that represent rhumb lines, or lines of constant compass bearing.
- **Transverse Mercator projection**: This is a cylindrical projection that is similar to the Mercator projection, but is oriented along a line of longitude rather than the equator. It is often used for mapping narrow strips of land, such as railway lines or pipelines.
- **Miller cylindrical projection**: This is a cylindrical projection that attempts to balance the distortions in both distance and area by stretching the map near the poles. It was developed by American cartographer Osborn Maitland Miller in 1942.

In summary, cylindrical projections are a type of map projection that projects the Earth's surface onto a cylinder, which is then unrolled to produce a flat map. They are commonly used for world maps and maps of large regions, because they preserve angles and shapes well and are relatively simple to produce and use. However, they have significant distortions in both distance and area, particularly at the poles.

```
fig, ax = plt.subplots(subplot_kw={'projection': ccrs.Mercator()})
fig.set_size_inches(10, 10)

countries.plot(ax=ax, color="black", transform=ccrs.PlateCarree())
```

```
geodata.plot(ax=ax, markersize=1, alpha=1, edgecolors='none',
             column="Maximum wind speed", cmap="jet",
             transform=ccrs.PlateCarree())
ax.gridlines(crs=ccrs.PlateCarree(), draw_labels=True,
                      linewidth=1, color='gray', alpha=0.8, linestyle='--')
plt.show()
```

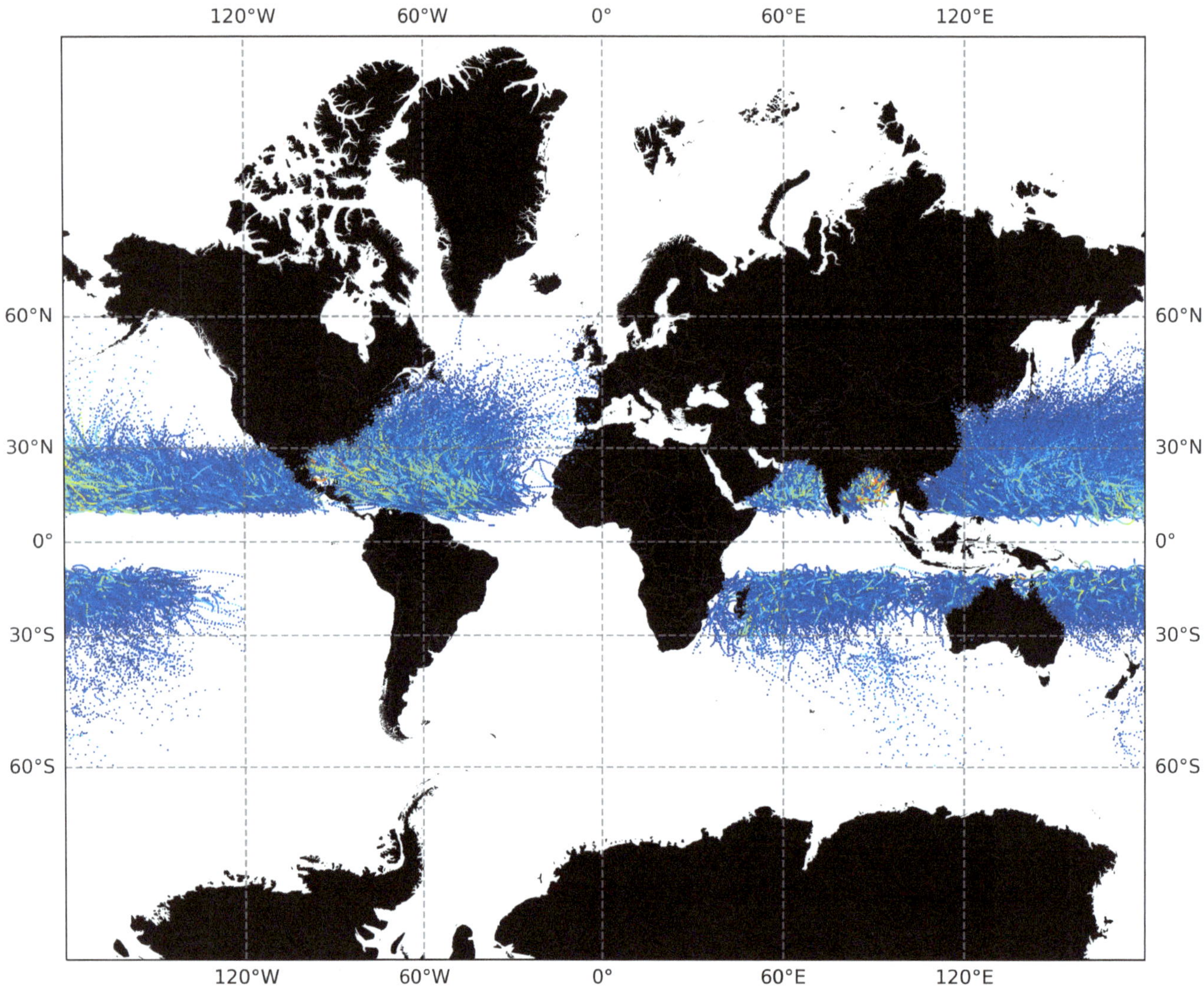

Figure 8.4: Tropical storms, plotted in the Mercator projection.

As we reach the conclusion of our exploration of map projections, we turn our focus to a projection that holds a special place in my cartographic preferences – the South Polar Stereographic projection. This projection offers a unique and insightful perspective, particularly apt for showcasing phenomena like tropical storms, which, while never crossing the equator, can have significant impacts in the southern hemisphere.

The South Polar Stereographic projection is a type of azimuthal projection centered on the South Pole. It provides an unparalleled view of the southern hemisphere, portraying areas near the pole with less distortion compared to many other projections. This characteristic makes it an excellent tool for examining weather patterns and natural events that predominantly affect the southern latitudes.

In this final section, we apply the South Polar Stereographic projection to our tropical storm data. This visualization will not only highlight the distinct advantages of this projection but also offer a fresh perspective on the data we have been analyzing. By focusing on the southern extremities of the Earth, we can gain new insights into the behavior and pathways of tropical storms in these regions.

This exercise serves as a reminder of the power of map projections to shape our understanding of geographical phenomena. Each projection tells a different story, and the South Polar Stereographic projection allows us to narrate a part of our planet's story that is often less emphasized in global maps.

As you view the final map, consider how the choice of projection can uncover new patterns and relationships, providing a deeper and more nuanced understanding of the data. This exploration of various projections, culminating in the South Polar Stereographic, underscores the importance of perspective in cartography and the role it plays in revealing the complexities of our world.

With this map, we conclude our journey through the diverse landscapes of map projections, having traversed the realms of planar, cylindrical, and conical types, each offering its own unique lens through which to view our planet and its phenomena.

```
fig, ax = plt.subplots(subplot_kw={'projection': ccrs.SouthPolarStereo()})
fig.set_size_inches(10, 10)

countries.plot(ax=ax, color="black", transform=ccrs.PlateCarree())
geodata.plot(ax=ax, markersize=1, alpha=1, edgecolors='none',
             column="Maximum wind speed", cmap="jet",
             transform=ccrs.PlateCarree())
ax.gridlines(crs=ccrs.PlateCarree(), draw_labels=True,
                      linewidth=1, color='gray', alpha=0.8, linestyle='--')
plt.show()
```

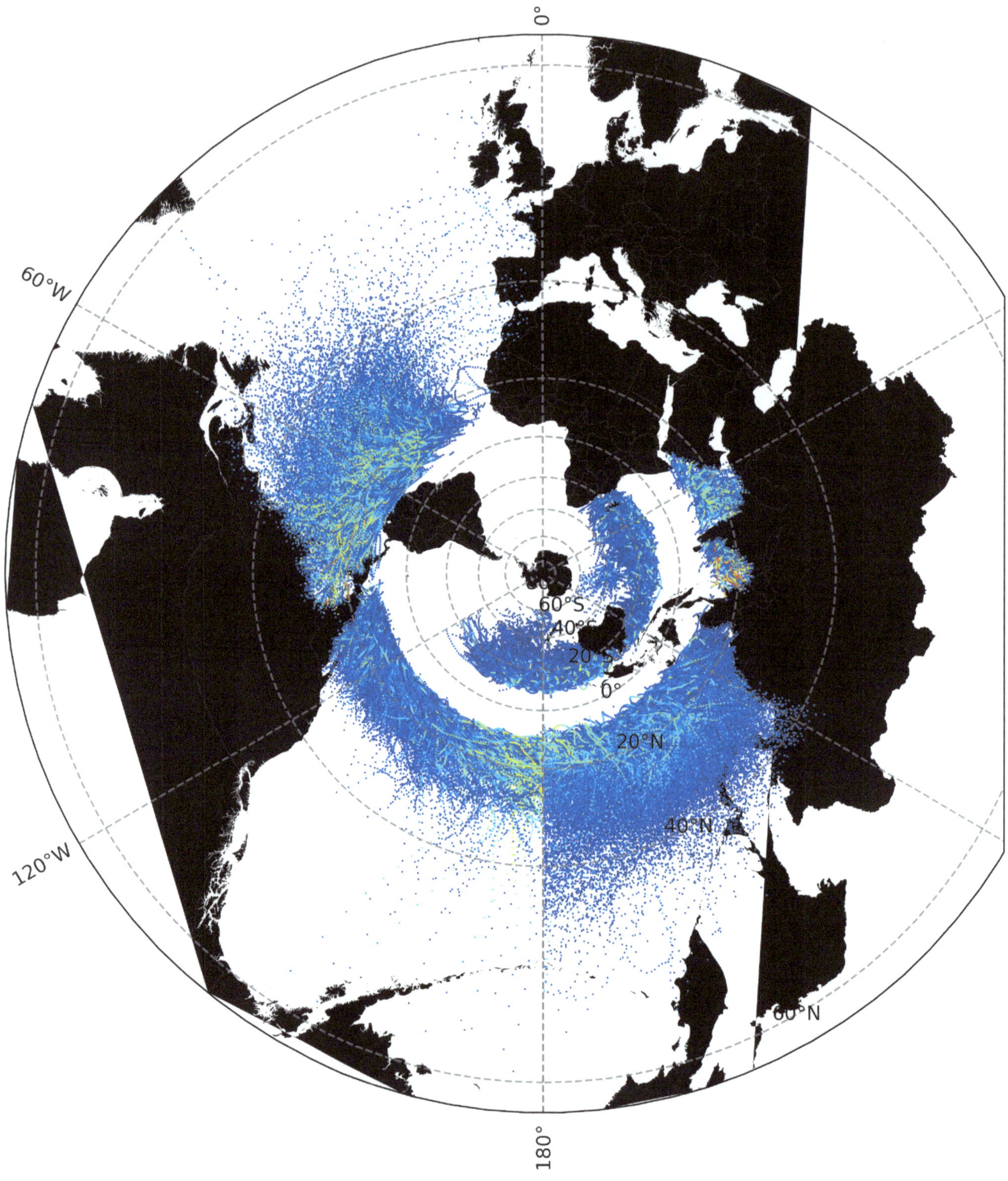

Figure 8.5: Tropical storms, plotted in the South Polar Stereographic projection.

9

Mapping Mars

9.1 Plotting Rasters - Mars Topography

In the fascinating journey of data visualization using Python, we have already navigated through the fundamentals of plotting raster data in our livestock chapter. As we continue our exploration, this chapter delves into the realm of advanced customization techniques that will elevate your raster plots from informative to insightful and visually compelling.

Our focus will be on three key areas. First, we will explore the art of adding text to your maps. This technique is not just about placing labels; it's about integrating textual data in a way that complements and enhances the visual narrative of your map. Whether it's annotating specific features or providing context, the strategic placement and styling of text can dramatically impact the readability and effectiveness of your plot.

Next, we will venture into the vibrant world of color maps. Color is a powerful tool in data visualization, capable of conveying moods, highlighting contrasts, and guiding the viewer's eye to key areas of interest. Customizing color maps goes beyond mere aesthetic choices; it involves understanding the psychology of colors and using it to underscore the differences and patterns in your data. We will learn how to tailor color maps to fit the story your data tells, ensuring that they are not just visually pleasing but also functionally significant.

Lastly, we will discuss the use of boundaries in controlling the application of color maps. This technique allows us to create more nuanced and detailed visual representations, especially when dealing with complex datasets. By setting boundaries, we can segment our data into more manageable parts, each highlighted with specific color schemes to denote different ranges or categories. This approach not only adds depth to our visualizations but also makes them more informative and easier to interpret.

To bring these concepts to life, we will use the intriguing example of a Mars topography map. This will serve as a practical case study to apply our advanced techniques, showcasing how text annotations, customized color maps, and the strategic use of boundaries can transform a simple topographic map into a rich, detailed, and informative visual story.

In this snippet, we are using the rasterio library, a powerful Python tool for working with raster data, which was introduced in the livestock chapter. Here, we're applying it to a different context — mapping the topography of Mars.

We start by importing rasterio. This library is essential for reading and writing raster data, which are grid-based datasets like images or 2D matrices. We then use the `rasterio.open()` function to open a raster file. In this case, the file is a Digital Elevation Model (DEM) of Mars, titled:
Mars_MGS_MOLA_DEM_mosaic_global_463m.tif.

This file likely contains elevation data for the Martian surface. Finally, we read the data from the file using the `read()` method. This loads the data into a variable (mars), which can then be used for further processing or visualization.

This code block uses numpy and the mars array (a raster dataset representing Martian topography) to analyze and print key statistics about Mars' surface elevations:

- **Importing numpy**: numpy is essential for numerical computations in Python.
- **Shape of the raster data**: `mars.shape` prints the dimensions (bands, rows, columns) of the Mars topography data, helping us understand its structure.
- **Minimum elevation**: `np.amin(mars[0])` calculates the lowest elevation point in the first band of the Mars data.
- **Maximum elevation**: `np.amax(mars[0])` finds the highest elevation point in the same band.
- **Total elevation range**: This expression calculates the total range

of elevation on Mars, from the lowest to the highest point: `np.amax(mars[0]) + abs(np.amin(mars[0]))`

This analysis not only sheds light on the Martian landscape but also illustrates a crucial aspect of raster data: they can represent real-world physical measurements (like elevation) or arbitrary scales (e.g., 0-256 for image intensities). This is key for correctly interpreting the data in various scientific and visualization contexts.

Mars has no oceans so there is no "sea level" reference point for altitude. "sea level" on Mars is defined as the average altitude on Mars. So negative topography values are areas below the average altitude and positive topography values are areas above the average altitude. Olympus Mons is the largest mountain on Mars, with a height of 21241m (relative to the average Martian altitude) and Hellas basin is the lowest point on Mars, with a depth of -8201m (relative to the average Martian altitude). Interestingly, this means that the difference between the highest and lowest points on Mars (~30 km) is greater than that on earth (~20km) between Mount Everest and the Marina trench.

```
import rasterio
import numpy as np

mars_file = rasterio.open('Mars_MGS_MOLA_DEM_mosaic_global_463m.tif')
mars = mars_file.read()

print(mars.shape)
print(np.amin(mars[0]))
print(np.amax(mars[0]))
print(np.amax(mars[0]) + abs(np.amin(mars[0])))
```

```
(1, 23040, 46080)
-8201
21241
29442
```

Using matplotlib and `rasterio.plot`, the code creates a visual representation of the Mars topography data. It employs a specific colormap ("viridis") for visual clarity and turns off the axis for a cleaner presentation, showcasing the advanced customization techniques in raster visualization.

Using the default colourmap reveals a problem, mapping the topography values to a colourmap linearly results in quite a bland map and most of the features are lost. The reason for this is that the majority of the Martian features lie somewhere between -8000–12000m. This should be evident from the average altitude being so skewed towards the Hellas basin and away from Olympus Mons. The four mountains which show up in yellow occupy a tiny portion of the planet (in terms of area) but because they are huge (in terms of altitude), a quarter of the colours in the colourmap exist just for them. So the rest of the features are mapped onto a small range of colours and clarity is poor.

```
import matplotlib.pyplot as plt
from rasterio.plot import show

fig, ax = plt.subplots()
fig.set_size_inches(14, 7)

show(mars_file, cmap="viridis")
ax.axis('off')

plt.show()
```

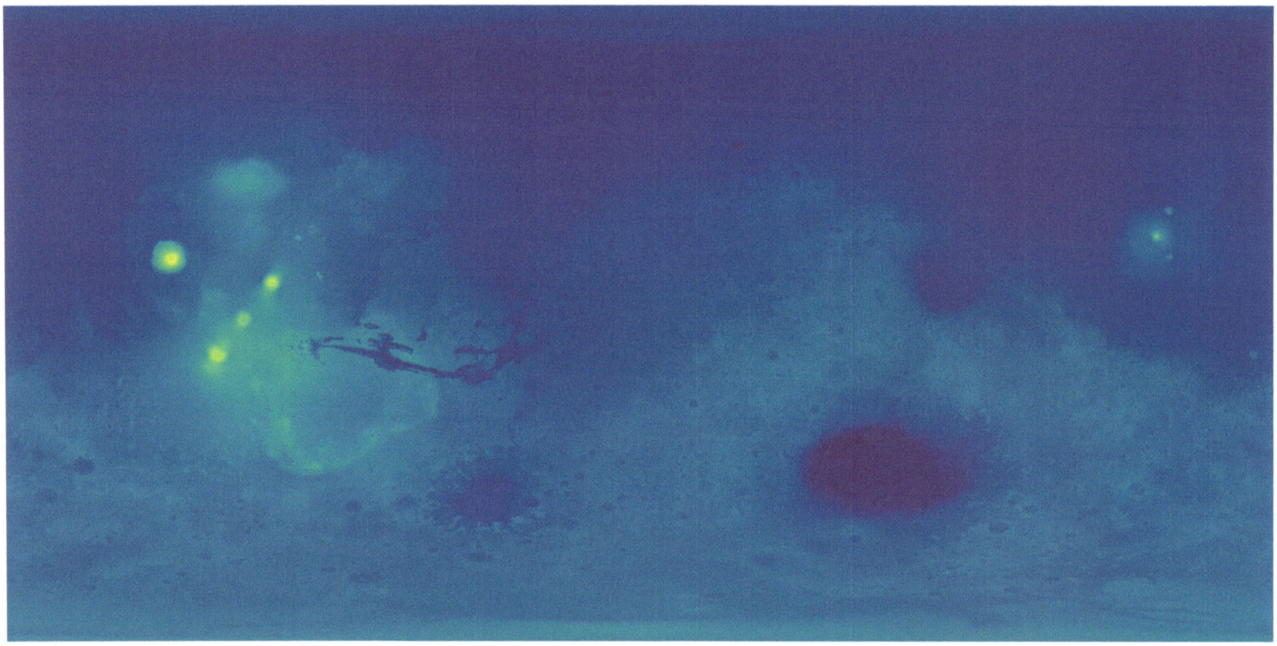

Figure 9.1: Martian topography.

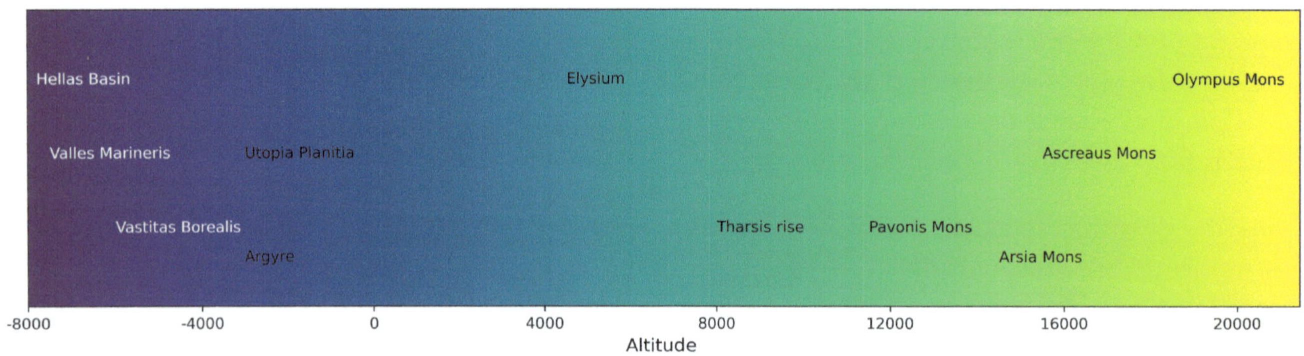

Figure 9.2: Relationship between colours in the viridis colourmap and Martian features.

To get around this we are going to create a colourmap and break some data visualisation rules. Strictly speaking we probably should have a linear colourmap normalised between the lowest and largest values in our data. However because in this case there are only four features above 14000m and a lot of features beneath 14000m, we are going to create a colourmap to map colours between -8201m (the lowest point in our data) and 14000m, with everything above 14000m being mapped to the last colour in the colourmap. Unfortunately matplotlib does not really have particularly good colourmaps to show topography. We are going to replicate the colourmap that NASA uses to show topography on Mars.

Below is a section of code that will create a colourmap loosely based on the one NASA uses, normalised between -8210m and 14000m in bands of 10m. The `LinearSegmentedColormap` method takes a list of colours and generates a colourmap with N colours spaced linearly between the colours in that list. The `BoundaryNorm` class allows you to map colours in a colourmap to a set of custom values. In this case we are creating a `BoundaryNorm` object which will map colours in bands of 10m between -8210 and 14000m. An example for clarity, values between 0–10m will be mapped to one colour, values between 10–20m will be mapped to the next colour and so on. Anything taller than 14000m will be coloured with the final colour in the colourmap.

Rainbow color maps, also known as "spectrum" or "jet" color maps, have been a popular choice in various fields for visualizing data. However, their use is increasingly debated, and in many cases, discouraged, due to several significant drawbacks:

- **Perceptual Non-Uniformity**: Rainbow color maps often create artificial boundaries and gradients that do not exist in the data. This happens because our perception of color changes does not linearly correspond to the actual numerical changes in data. For instance, certain colors in the spectrum (like yellow and cyan) may appear to jump out, implying a more significant change in data than there really is.
- **Color Vision Deficiency (CVD) Accessibility**: A significant portion of the population has some form of CVD (commonly known as color blindness). Rainbow color maps can be particularly challenging for these individuals, as they often rely on color distinctions that are difficult or impossible for people with CVD to discern. This can lead to misinterpretation or complete inability to interpret the data accurately.

- **Loss of Quantitative Information**: In a rainbow color map, the colors don't have an intuitive order or scale (unlike, for example, a grayscale or a single-hue gradient), which can make it hard to estimate quantitative values from the colors. This can lead to misinterpretations of the data's actual values.
- **Printing and Grayscale Conversion Issues**: Rainbow color maps can lose their effectiveness when converted to grayscale, which is common in printed documents. The different colors may end up having similar shades of gray, making it hard to distinguish the data in the absence of color.

Despite these drawbacks, rainbow color maps may still be used in certain contexts, primarily for historical reasons or because they can be visually striking and engage viewers. However, the trend in data visualization is increasingly towards more perceptually-linear color maps (like "viridis" or "plasma" in matplotlib), which provide a more accurate and accessible representation of the data.

```
from matplotlib.colors import BoundaryNorm, LinearSegmentedColormap
import numpy as np

custom_cmap = LinearSegmentedColormap.from_list('mars', ['#162252',
                                                          '#104E8B',
                                                          '#00B2EE',
                                                          '#00FF00',
                                                          '#FFFF00',
                                                          '#FFA500',
                                                          '#FF0000',
                                                          '#8b0000',
                                                          '#964B00',
                                                          '#808080',
                                                          '#FFFFFF'], N=2221)

bounds = np.arange(-8210, 14000, 10)
norm = BoundaryNorm(bounds, custom_cmap.N)
```

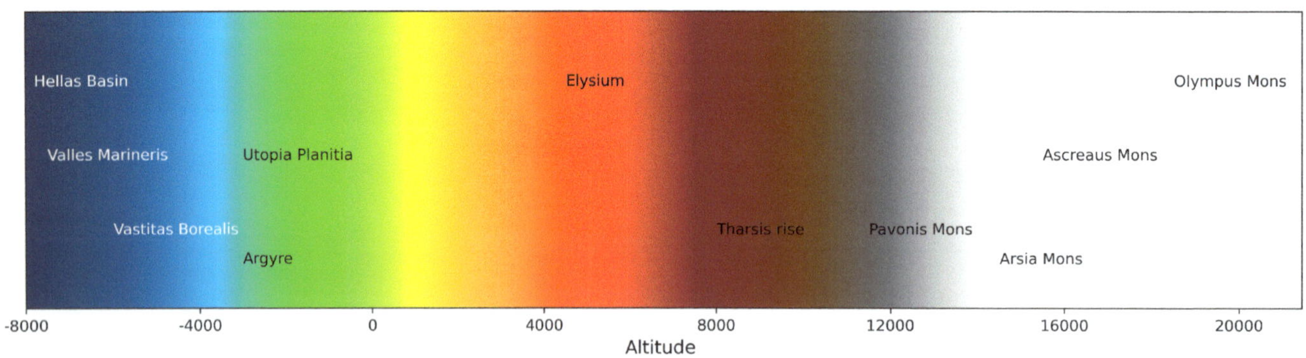

Figure 9.3: Relationship between the colours in our custom colourmap and Martian features.

The Martian topography map can now be replotted with the new colourmap and the values normalised. It is quite amazing how different two plots can be simply by changing the colourmap and tweaking the way colours are mapped to values.

```
fig, ax = plt.subplots()
fig.set_size_inches(14, 7)

show(mars_file, cmap=custom_cmap, norm=norm)
ax.axis('off')
plt.show()
```

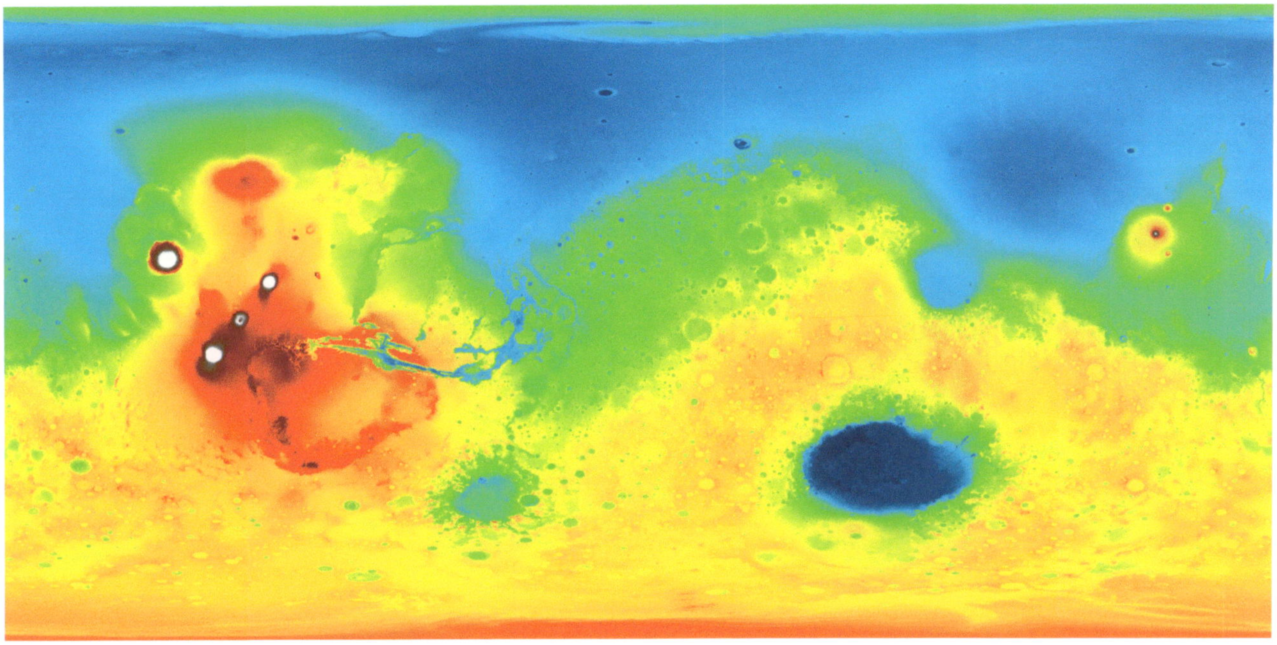

Figure 9.4: Martian topography with our custom colourmap.

When creating a topography map, especially one as complex as that of Mars, adding labels to specific features can greatly enhance the understanding and usefulness of the map. The `plt.text` function in Matplotlib is a straightforward and effective way to add such annotations.

`plt.text` is used for placing text on a plot. It's particularly useful for labeling specific data points, regions, or features of interest in a plot. The function typically requires coordinates for placement (x, y) and the text string you want to display. For example, `plt.text(x, y, "label")` would place the text "label" at the coordinates (x, y) on the plot. `plt.text` also offers various customization options, such as adjusting the font size, style, color, and alignment of the text. This allows you to make the text easily readable and ensures that it blends well with the rest of your plot without obscuring important information. When labeling features on a Mars topography map, you'd use `plt.text` to annotate notable landmarks, such as craters, mountains, valleys, or regions of scientific interest. By providing these labels, you can make the map more informative and easier to navigate, especially for readers who may not be familiar with Martian geography. One challenge with adding text to maps is avoiding overlap with other plot elements. Strategic placement and possibly adjusting text size or using text boxes with background color can help make sure that each label is clear and legible.

```
fig, ax = plt.subplots()
fig.set_size_inches(14, 7)

show(mars_file, cmap=custom_cmap, norm=norm)
ax.text() #x, y, text as many times as you like
ax.axis('off')
plt.show()
```

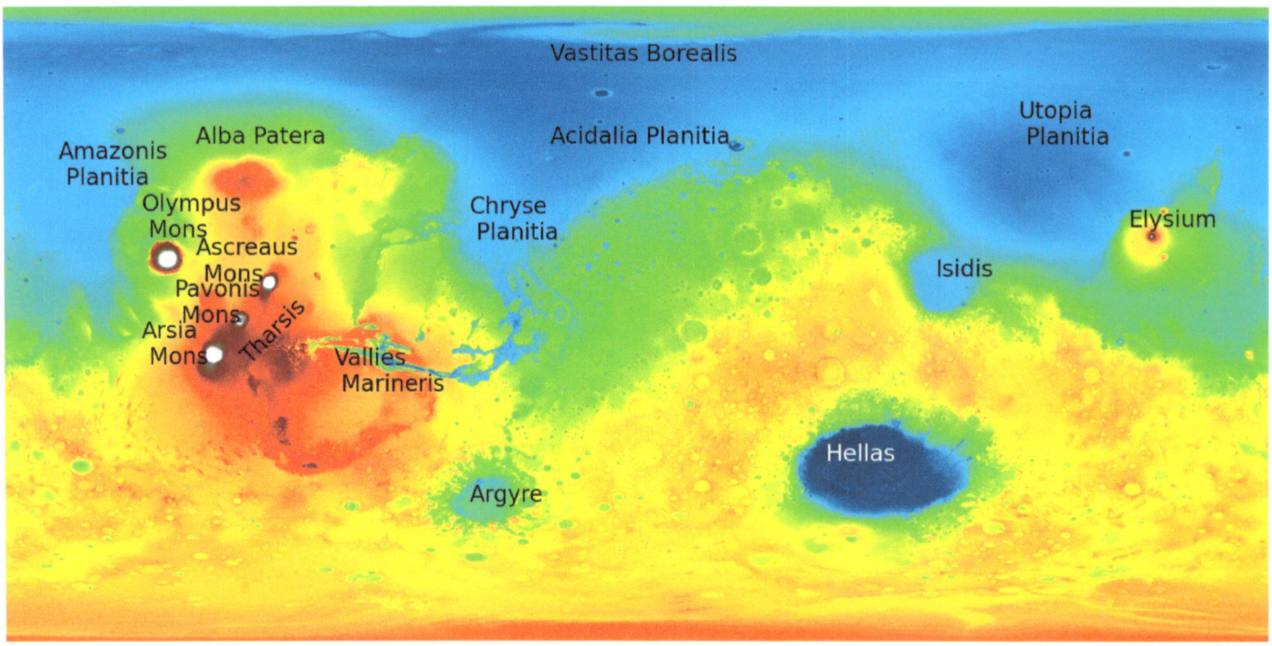

Figure 9.5: Martian topography with labelled features.

In the next section, we'll explore the concept of visualizing a terraformed, wet Mars by creating an "Earth-like" color map in Matplotlib. The term "Earth-like" here refers to a color map that uses deep to light blue shades for elevations below sea level and transitions to green to brown tones for elevations above sea level, mimicking the Earth's diverse landscapes.

While Matplotlib offers a 'terrain' color map designed for such purposes, it falls short in representing the deep blues of Earth's oceans. To overcome this, we'll ingeniously combine the 'terrain' color map with the 'ocean' color map, enhancing the blue hues for a more realistic portrayal of sub-sea-level topography.

The process involves manipulating and merging color maps. Color maps in Matplotlib are essentially arrays of colors ranging from 0 to 1. We can extract a subset of these colors to tailor the color maps to our needs. For instance, `plt.cm.ocean(np.linspace(0.2, 0.8, 821))` fetches 821 colors from the 'ocean' color map, while deliberately omitting colors below the 0.2 and above the 0.8 mark. These range from medium to light blues. Similarly, we extract 1400 colors from the 'terrain' color map. These cover a spectrum of earthy tones. These extracted color segments are then seamlessly merged using the `LinearSegmentedColormap` method, creating our custom Earth-like color map.

The choice of 821 colors from 'ocean' and 1400 from 'terrain' isn't arbitrary. It's intricately linked to how we wish to map topographical data to colors. The BoundaryNorm class in Matplotlib allows us to map specific ranges of data values to certain colors in our color map. In our case, we create a color map with 2221 distinct colors (821 from 'ocean' plus 1400 from 'terrain') and use `BoundaryNorm` to assign these colors to corresponding topographical values. The levels are set such that the first 821 levels (representing elevations below sea level) are mapped to the 'ocean' colors, while the next 1400 levels (for elevations above sea level) are mapped to the 'terrain' colors. This results in a nuanced and visually appealing representation of Martian topography as if it were Earth, with distinct colorations for underwater and above-ground features.

This approach provides a nuanced color representation of Mars' topography, distinguishing underwater and above-ground features as if the planet were Earth-like, thereby enhancing the visual appeal and informational value of the topographical data.

```
colors_undersea = plt.cm.ocean(np.linspace(0.2, 0.8, 821))
undersea_map = LinearSegmentedColormap.from_list('undersea_map', colors_undersea, N=821)

colors_land = plt.cm.terrain(np.linspace(0.25, 1, 1400))
land_map = LinearSegmentedColormap.from_list('land_map', colors_land, N=1400)

colors = np.vstack((colors_undersea, colors_land))
terrain_map = LinearSegmentedColormap.from_list('cut_terrain', colors, N=2221)

bounds = np.arange(-8210, 14000, 10)
norm = BoundaryNorm(bounds, terrain_map.N)
```

Plotting the data using this new colourmap and colour boundaries gives an almost earth like planet.

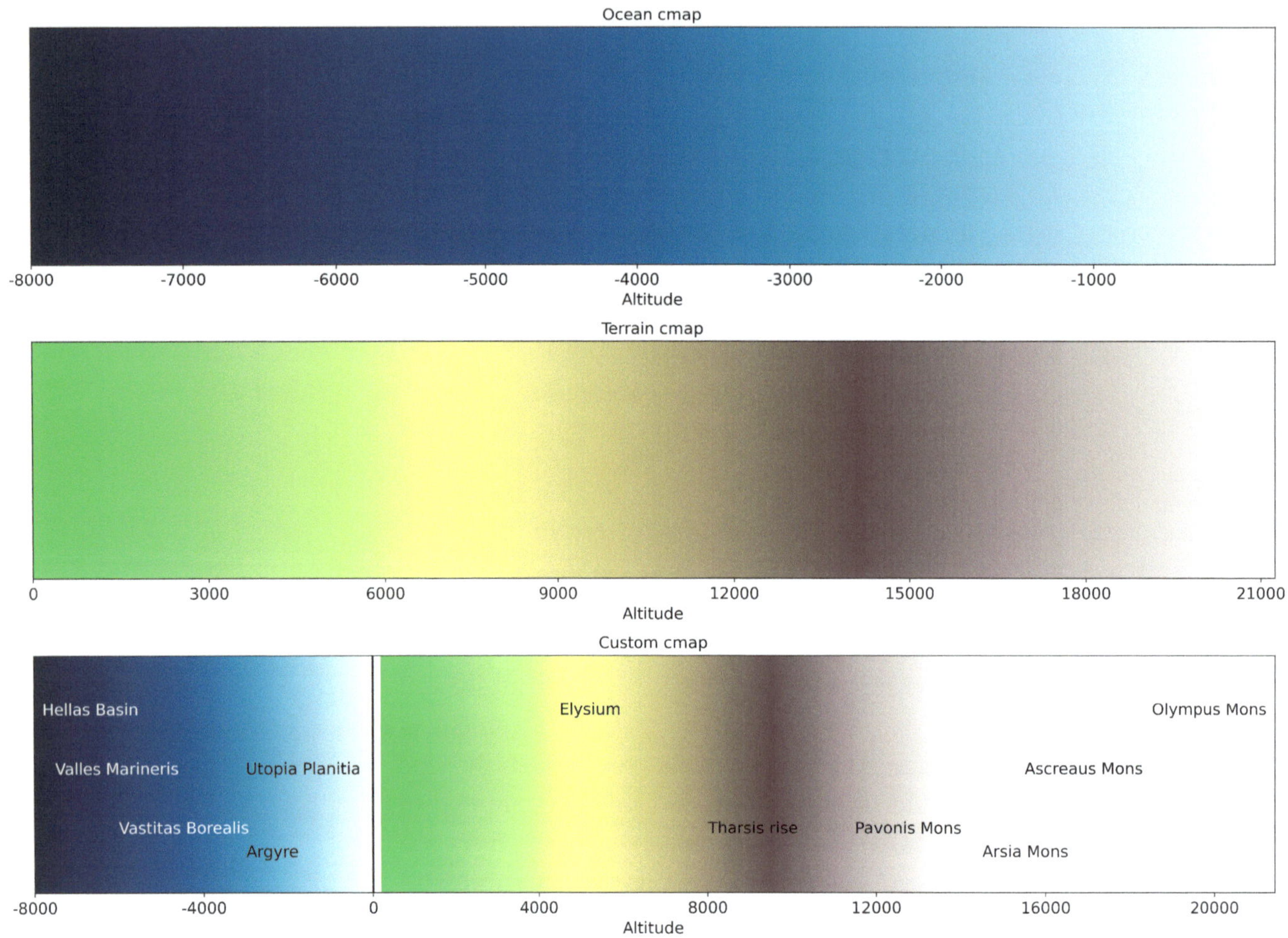

Figure 9.6: Relationship between Martian features and our custom colourmap.

```

fig, ax = plt.subplots()
fig.set_size_inches(14, 7)

show(mars_file, cmap=terrain_map, norm=norm)
ax.text()  #x, y, text as many times as you like
ax.axis('off')
plt.show()
```

We have chosen an arbitrary value for our hypothetical sea level by just using the average altitude of Mars as sea level. However using the method for generating colourmaps outlined above, it is possible to change where the sea level is. Recall, we generate our colourmap of 2221 colours, with 821 coming from the ocean colourmap and 1400 coming from the terrain colourmap. By changing the number of colours from each colourmap, without changing the total number of colours in the overall colourmap it is possible to change where our sea level is. Or in simpler terms, where the blue stops and the green starts.

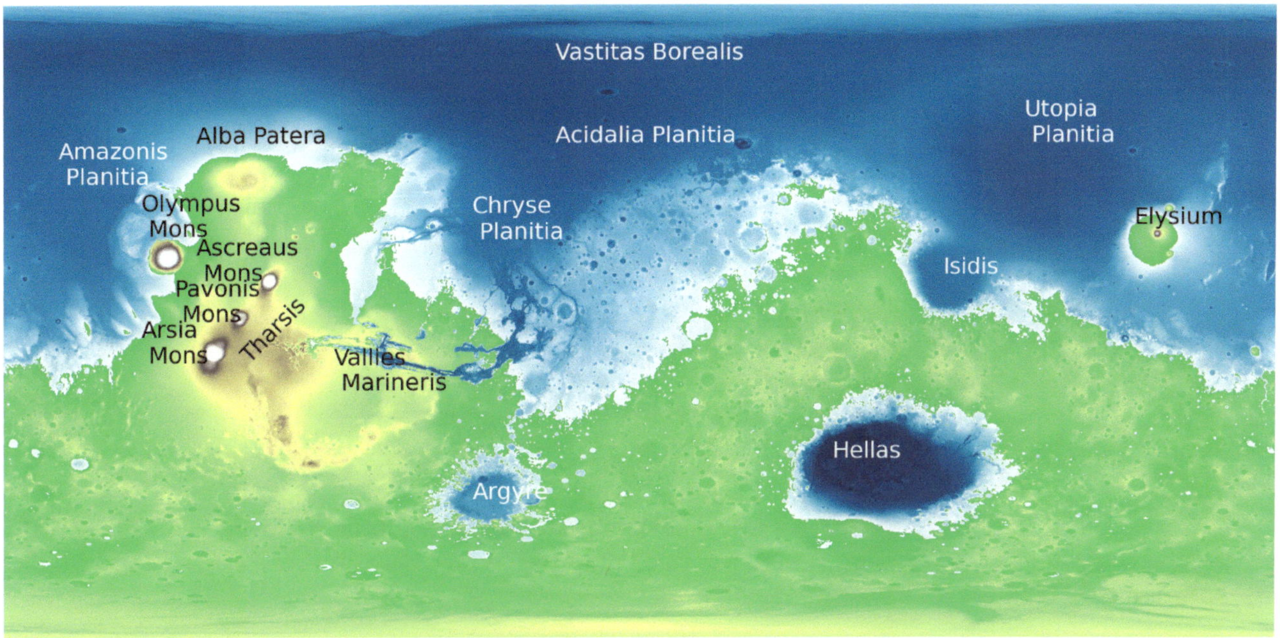

Figure 9.7: Martian topography plotted with an "Earth" like colourmap.

Below is a function to take a sea level value and generate a colourmap and colour boundary object to be used in plotting. This code block is designed to create a custom color map for visualizing Martian topography, taking into account a hypothetical new sea level. It involves adjusting the proportion of undersea and land colors from Matplotlib's 'ocean' and 'terrain' color maps to match this new sea level. Here's a breakdown of the code:

`new_sea_level = 1000` sets the new sea level to 1000 meters.

`new_sea_level = new_sea_level / 10`: The sea level is then divided by 10. This step seems to be normalizing the sea level value, possibly to match the scale of the color map or elevation data.

`undersea = int(800 + new_sea_level)` calculates the number of undersea colors to include. It starts with a base of 800 and adds the normalized new sea level.

`land = int(1400 - new_sea_level)` determines the number of land colors, starting from 1400 and subtracting the normalized new sea level. This ensures the total number of colors (undersea + land) aligns with the predefined total in the color map.

`colors_undersea = plt.cm.ocean(np.linspace(0.2, 0.8, undersea))` extracts a specific range of colors from the 'ocean' color map based on the calculated undersea value.

`undersea_map = LinearSegmentedColormap.from_list(...)`: Creates a new segmented color map for undersea colors.

`colors_land = plt.cm.terrain(np.linspace(0.25, 1, land))`: Similarly extracts land colors from the 'terrain' color map.

`land_map = LinearSegmentedColormap.from_list(...)`: Generates a segmented color map for land colors.

`colors = np.vstack((colors_undersea, colors_land))`: stacks the undersea and land colors to form a unified array.

`terrain_map = LinearSegmentedColormap.from_list(...)`: creates a comprehensive color map named 'cut_terrain' with these combined colors.

`bounds = np.arange(-8210, 14000, 10)`: establishes an array of boundary values for elevation levels.

`norm = BoundaryNorm(bounds, terrain_map.N)`: Uses BoundaryNorm to map elevation values to the appropriate colors in the 'cut_terrain' color map.

```
def change_sea_level(new_sea_level):
    new_sea_level = new_sea_level / 10
    undersea = int(800 + new_sea_level)
    land = int(1400 - new_sea_level)
    colors_undersea = plt.cm.ocean(np.linspace(0.2, 0.8, undersea))
    undersea_map = LinearSegmentedColormap.from_list('undersea_map', colors_undersea,
                                                     N=undersea)

    colors_land = plt.cm.terrain(np.linspace(0.25, 1, land))
    land_map = LinearSegmentedColormap.from_list('land_map', colors_land, N=land)

    colors = np.vstack((colors_undersea, colors_land))
    terrain_map = LinearSegmentedColormap.from_list('cut_terrain', colors, N=2221)

    bounds = np.arange(-8210, 14000, 10)
    norm = BoundaryNorm(bounds, terrain_map.N)
    return terrain_map, norm


terrain_map0, norm0 = change_sea_level(new_sea_level = 0)
terrain_map3000, norm3000 = change_sea_level(new_sea_level = 3000)
terrain_map_3000, norm_3000 = change_sea_level(new_sea_level = -3000)
```

This can then be put to use. Below is an example of what Mars would look like if the sea level was set to being 1500m above the average altitude. Which vaguely corresponds to Mars with a 70% coverage of

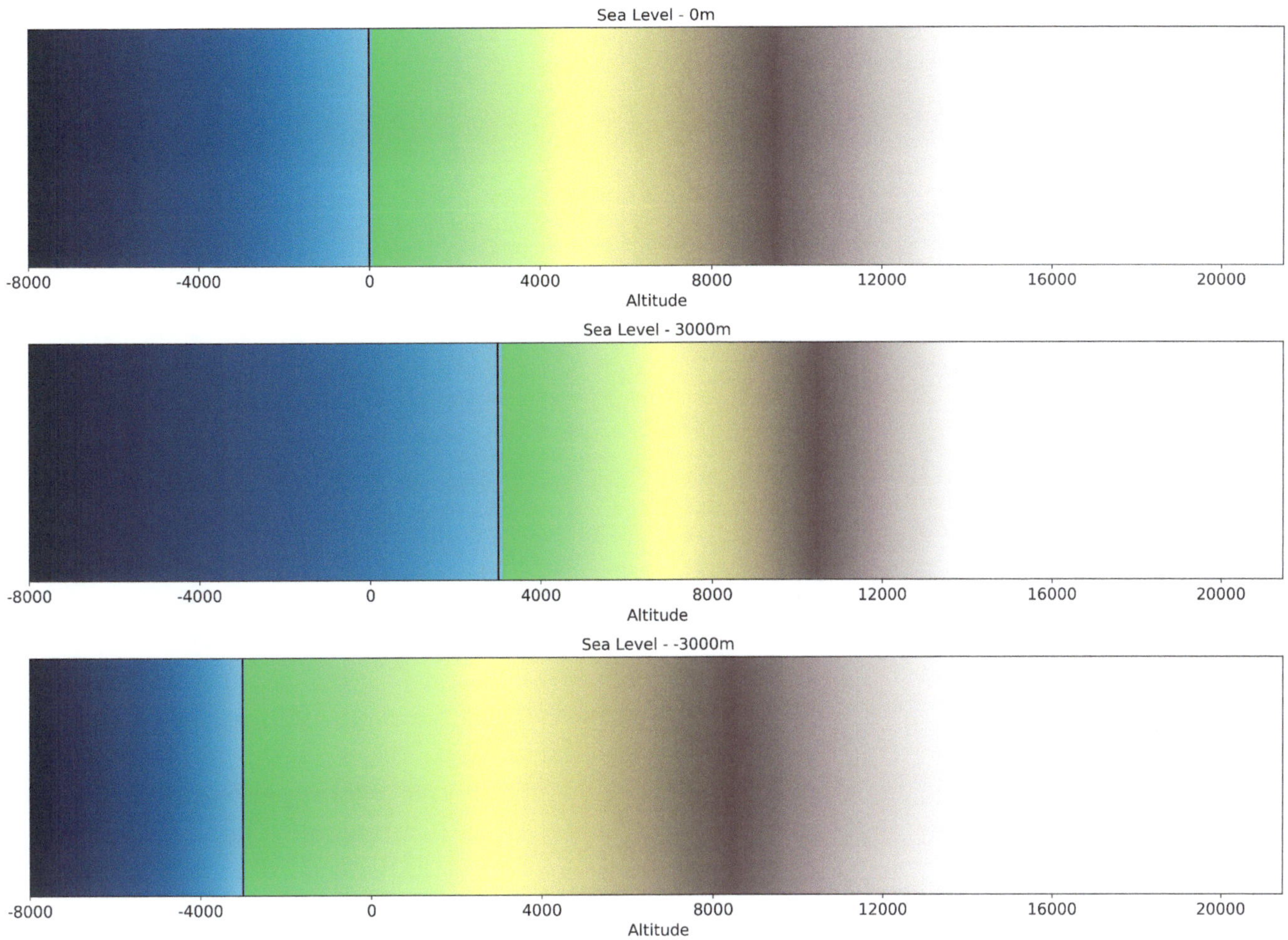

Figure 9.8: Custom "Earth" like colourmap with different sea levels.

water, which is what we have on Earth.

```
terrain_map1500, norm1500 = change_sea_level(new_sea_level = 1500)

fig, ax = plt.subplots()
fig.set_size_inches(14, 7)

show(mars_file, cmap=terrain_map1500, norm=norm1500)
ax.text()  #x, y, text as many times as you like
ax.axis('off')
plt.show()
```

In this section, you are going to enhance the visualization of Martian topography by adding hillshade. Hillshade is a method used to create a three-dimensional effect on a two-dimensional map, making it more visually appealing and informative. Hillshading will be covered in more detail in the next chapter on topography so don't worry about the sparse detail here.

Hillshade simulates the effect of light falling on the topography, creat-

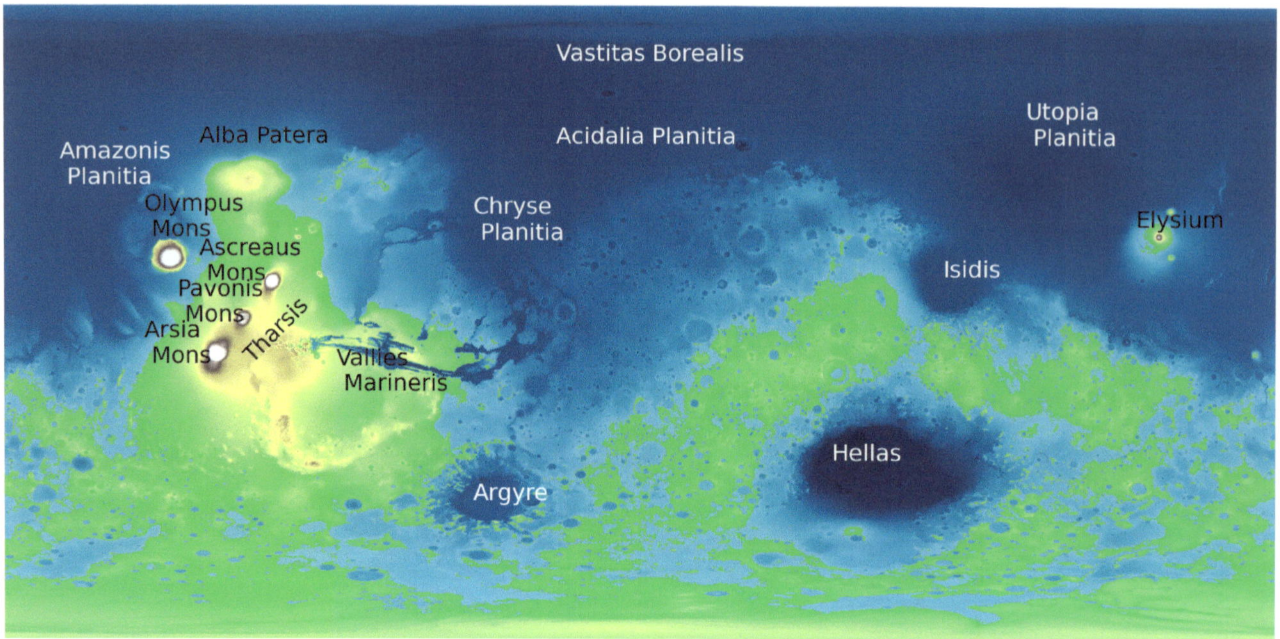

Figure 9.9: Martian topography with an elevated sea level.

ing shadows and highlights that mimic a 3D appearance. It's typically rendered in grayscale, where darker and lighter colors represent shadows and highlights, respectively. Hillshade is commonly used as an underlay in maps to give a more three-dimensional and realistic feel to the terrain representation. This can significantly enhance the visual impact and interpretability of the map. The earthpy library's hillshade function is a convenient tool to generate hillshade data. It's particularly user-friendly and well-suited for this kind of topographical visualization.

- Tunable Parameters:
  - **Azimuth**: This parameter, ranging from 0 to 360 degrees, controls the direction from which the light source appears to shine. An azimuth of 0 degrees means the light is coming from the north. Adjusting the azimuth can dramatically change the appearance of the hillshade, as it alters the direction of shadows.
  - **Altitude**: The altitude parameter, ranging from 1 to 90 degrees, represents the angle of the light source above the horizon. A higher value simulates a light source that is higher in the sky, affecting the length and angle of shadows.

The choice of azimuth and altitude values can significantly influence the final appearance of the hillshade and should be tuned based on the specific characteristics of the dataset and the desired visual effect.

We can plot the data in the same manner as other two dimensional arrays using the *rasterio.show.plot* method. The only consideration is that we need to set the transform value to be equal to the transform of the original file.

```
import earthpy as et
import earthpy.spatial as es

hillshade = es.hillshade(mars[0], azimuth=250, altitude=1)
fig, ax = plt.subplots(facecolor='#FCF6F5FF')
fig.set_size_inches(14, 7)
show(mars_file, cmap=terrain_map, norm=norm)
show(hillshade, transfor=mars_file.transform, cmap="Greys", alpha=0.2)
ax.text()
ax.axis('off')
plt.show()
```

Figure 9.10: Martian topography with hillshade.

10

Mapping Topography

10.1 Merging and Cutting Rasters - Topography

In this chapter, we delve into the intricate world of advanced raster operations by creating a topographic map of Pakistan. The journey into advanced raster techniques is a continuation of our exploration of raster data, building upon the foundational knowledge previously established.

We begin by exploring the concept of raster merging or mosaicking. This technique is essential for creating a cohesive and expansive representation of geographical areas, particularly large or complex regions like Pakistan, which may not be adequately covered by a single raster dataset. Through practical examples, we will guide you through the process of merging multiple raster datasets, employing rasterio.

Moving forward, we will introduce the technique of masking rasters using polygons. This method is pivotal in isolating specific areas of interest within a larger dataset. We will demonstrate how to use a polygon – in this case, the geographical boundary of Pakistan – to mask out extraneous data, thereby concentrating our focus on the relevant topographical features. This section will include hands-on examples using tools like GDAL, rasterio, or Geopandas, illustrating how to efficiently apply this technique to our topographic map.

Further enhancing our map's visual appeal and interpretability, we will revisit the application of hillshade. While hillshade was discussed in the previous chapter, here we will tailor its application to the topography of Pakistan. Hillshade will bring a three-dimensional quality to the map, accentuating the ruggedness and intricacies of Pakistan's terrain. We'll provide practical insights into generating hillshade using tools such as EarthPy, highlighting how this technique can dramatically enhance the understanding and aesthetic of topographic maps.

Our culmination point will be a case study where we combine all these techniques to create a detailed topographic map of Pakistan. Starting from acquiring the necessary raster data, we will walk through the processes of merging, masking, and applying hillshade. Special attention will be given to the selection of parameters and methods that best suit the unique topographical characteristics of Pakistan.

Merging Rasters

We're using Python's rasterio library to merge multiple raster datasets into a single, cohesive raster image. This technique is particularly valuable when dealing with topographical data that is spread across several files, as it allows for the creation of a comprehensive and continuous map from these disparate pieces.

The code defines a list named *file_list*, which contains the file paths of the raster datasets to be merged. These files represent different sections of topography, likely covering various geographical regions. In practical applications, these files might represent different tiles of a larger area.

Following this, the code uses a list comprehension to open each raster file specified in file_list. This is achieved using the `rasterio.open` function, which returns a `DatasetReader` object for each file. These objects are stored in a list named *srcs_to_mosaic*, which essentially becomes a collection of the raster datasets ready to be merged.

The core of the process is the `merge` function, which takes *srcs_to_mosaic* as its input. This function seamlessly combines the input rasters into a single array. The resulting output is twofold: *arr*, a NumPy array that holds the merged raster data, and *out_trans*, the corresponding output transform. In the context of raster data, a transform is crucial as it maps the row and column coordinates of the raster to spatial coordinates, which is essential for accurate geographical representation. The end product of this operation is a unified raster dataset represented by *arr*, merging the topographical information from all the input files.

```python
import rasterio
from rasterio.merge import merge

file_list = ['10N060E_20101117_gmted_mea075.tif',
             '30N060E_20101117_gmted_mea075.tif',
```

```
             '10N030E_20101117_gmted_mea075.tif',
             '30N030E_20101117_gmted_mea075.tif']

srcs_to_mosaic = [rasterio.open(f) for f in file_list]
arr, out_trans = merge(srcs_to_mosaic)
```

In this code snippet, we are using the show function from `rasterio.plot` to visualize the merged raster data (*arr*) obtained from the previous merging process. The key aspect here, which hasn't been detailed before, is the use of the transform argument within the show function. Let's delve into what transform means and its significance in the context of displaying raster data.

The transform argument in rasterio and its role in the show function is crucial for accurately positioning and scaling raster data in a spatial context. A transform in raster data is essentially a set of coefficients that maps pixel coordinates in the raster (row and column indices) to spatial coordinates in a geographic coordinate system. This mapping is vital for any geospatial analysis, as it ensures that the raster data is correctly aligned with its geographical location on the Earth's surface.

In the code, *out_trans* is the transform associated with the merged raster dataset (*arr*). When we merged multiple raster datasets in the previous step, we not only combined their data but also needed to reconcile their spatial references into a single, cohesive transform. This output transform (*out_trans*) represents this cohesive spatial reference.

By passing *out_trans* to the show function, we instruct rasterio to use this spatial information to correctly display the raster. This ensures that the raster is properly oriented and scaled according to its real-world geographical coordinates. Without the correct transform, the raster data could be misrepresented, leading to incorrect interpretations.

```
import matplotlib.pyplot as plt
from rasterio.plot import show

show(arr, transform=out_trans)
```

Masking Rasters

In this code snippet, you are creating a memory file to temporarily store the merged raster data (*arr*) in memory, a technique particularly useful for intermediate processing steps such as clipping the data to a specific region. Let's explore what memory files are and the significance of each component of this code.

A memory file in the context of raster data processing with rasterio is

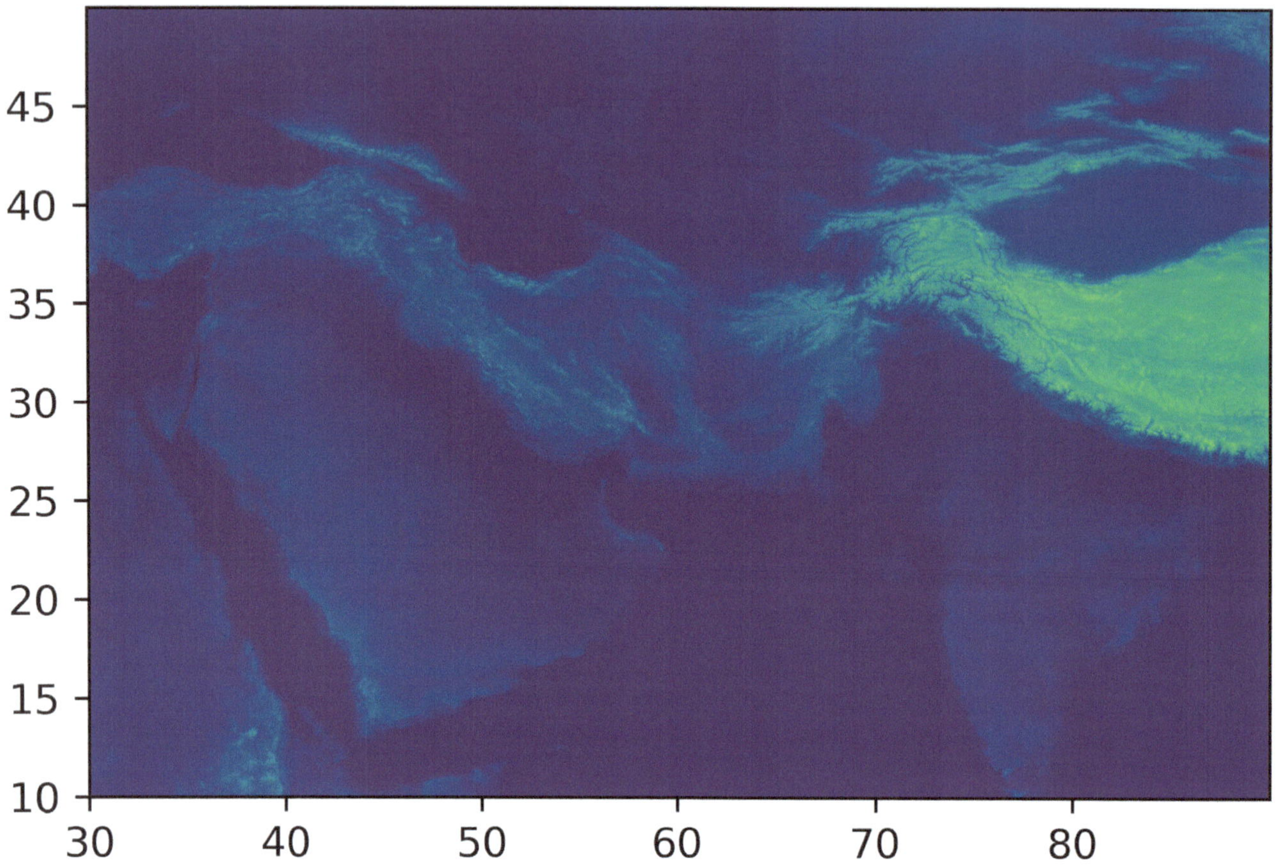

Figure 10.1: Topographical map of Central Asia, built from four merged TIFF files.

essentially a temporary file that exists only in your computer's memory (RAM), not on the disk. This approach is beneficial for several reasons:

- Speed: Operations on data stored in memory are typically faster than those performed on disk-based files. This is because accessing and writing data in RAM is quicker than doing so on a hard drive.
- No Physical File Creation: Memory files avoid the need to create and manage physical files on the disk, which is especially useful for intermediate steps in a data processing workflow where the output doesn't need to be permanently stored.
- Ease of Cleanup: Since memory files exist only during the runtime of the program, there's no need for manual cleanup (deletion) of temporary files, simplifying the code and reducing the risk of leaving behind unused data.

`MemoryFile()` initializes a new memory file and `memfile.open(...)`

opens a new dataset in the memory file. The parameters define the characteristics of this dataset: *driver='GTiff'* specifies the format of the dataset (GeoTIFF in this case). height and width set the dimensions of the dataset to match those of the first band of the merged raster data (arr[0]). count=1: Indicates that the dataset will contain a single band. crs: Sets the coordinate reference system to match that of the first source raster in the mosaic. *transform=out_trans* ensures that the spatial referencing (transform) of the merged data is retained. *dtype=arr.dtype* sets the data type to match that of the merged raster data.
`dataset.write(arr[0], 1)` This writes the first band of the merged raster data (*arr[0]*) into the newly created dataset in the memory file. The 1 indicates that this data should be written to the first band of the dataset.

By using a memory file, you're setting up a temporary and efficient storage mechanism for your merged raster data, preparing it for subsequent operations like clipping to the boundaries of Pakistan. This approach streamlines the workflow and enhances performance, especially when dealing with large datasets or complex processing tasks.

```
from rasterio.io import MemoryFile

memfile = MemoryFile()
dataset = memfile.open(driver='GTiff', height=arr[0].shape[0],
                       width=arr[0].shape[1], count=1, crs=srcs_to_mosaic[0].crs,
                       transform=out_trans, dtype=arr.dtype)
dataset.write(arr[0], 1)
```

In this next code snippet, we're employing the mask function from the `rasterio.mask` module to clip the merged raster data to the shape of Pakistan. This operation involves using a polygon that defines the geographical boundaries of Pakistan to extract only the part of the raster data that falls within these boundaries. Let's delve into the details of this process:

`clipped_array, clipped_transform = mask(...)` is where the actual clipping occurs. The mask function takes several arguments: *dataset* which is the raster dataset to be clipped (the merged topography data stored in a memory file).
`[mapping(geom) for geom in (pakistan.geometry.tolist())]` converts the geometry of Pakistan (extracted from the Geopandas dataframe) into a format suitable for rasterio. Essentially, it's a list of geometries in the correct format. *crop=True* specifies that the output raster should be cropped to the extent of the mask (Pakistan's boundary). *nodata=20000* sets a nodata value for the operation, which is used to fill the parts

of the raster that fall outside the mask. `clipped_array[clipped_array == 20000] = np.amin(clipped_array) - 1` replaces these nodata values with a value that is one less than the minimum value in the array. This step is often done for visualization purposes to clearly distinguish no-data areas.

In essence, this process involves using a vector shape (the boundary of Pakistan) to clip or extract a portion of a raster dataset. The result, clipped_array, is a new raster where only the area within the boundaries of Pakistan is retained, and everything outside these boundaries is set to a specified nodata value.

```
import geopandas as gpd
import numpy as np

from shapely.geometry import mapping
from rasterio.mask import mask

df = gpd.read_file('NaturalEarth/10m_cultural/ne_10m_admin_0_countries.shp')
pakistan = df.loc[df['ADMIN'] == 'Pakistan']

clipped_array, clipped_transform = mask(
    dataset, [mapping(geom) for geom in (pakistan.geometry.tolist())],
    crop=True, nodata=20000)

clipped_array[clipped_array == 20000] = np.amin(clipped_array) - 1
```

```
import matplotlib.pyplot as plt

fig, ax = plt.subplots()
show(clipped_array, ax=ax, transform=clipped_transform)
plt.show()
```

By utilizing the `LinearSegmentedColormap.from_list` method from Matplotlib's colors module, we define a new color map named 'pakistan' that linearly transitions between the colors black, green, and white – which are the primary colors of Pakistan's flag. This color map can then be applied to your raster plot, providing a unique and thematic visual representation that resonates with the subject of your map, Pakistan's topography. This approach not only adds a layer of customization to your visualization but also embeds a national symbolic reference into the data representation.

Next, we want to customize the color map and normalization for your topography plot of Pakistan.

`newcolors = cmap(np.linspace(0, 1, int((abs(np.amin(clipped_array) + abs(np.amax(clipped_array)))) / 10)))` generates a range of colors from the previously defined 'Pakistan' color map (cmap) based on the elevation range of your clipped raster data (clipped_array). The

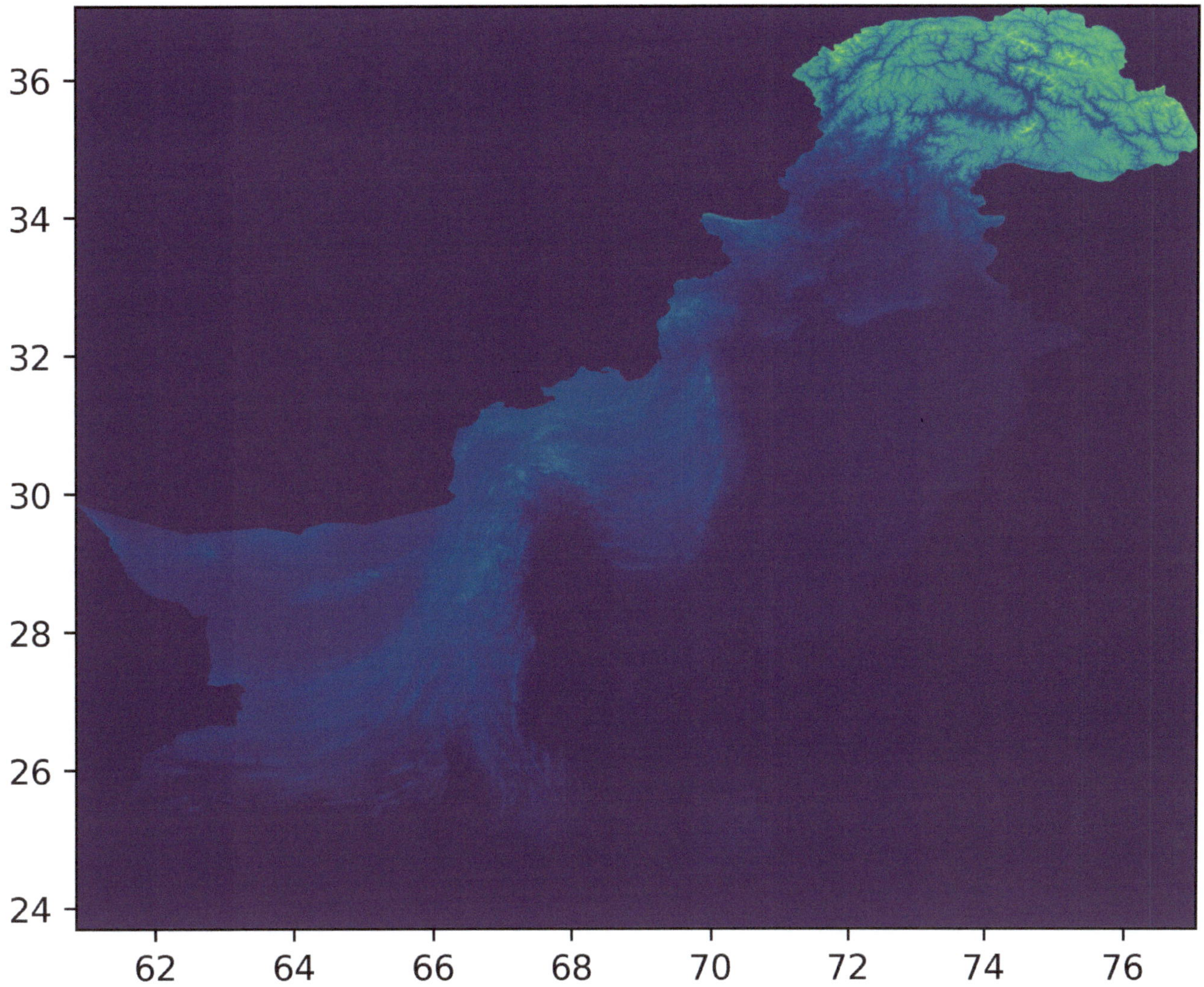

Figure 10.2: Topographical map of Pakistan.

range is divided into intervals of 10 meters each, from the minimum to the maximum elevation. `newcolors[:1, :] = np.array([1.0, 1.0, 1.0, 1.0])` sets the first row of the color map to white. `cmap = ListedColormap(newcolors)` creates a new `ListedColormap` object from the array of colors (newcolors). This color map is now a discrete representation of the elevation bands.

`bounds = np.arange(np.amin(clipped_array), np.amax(clipped_array) -4000, 10)` sets up the boundaries for normalization. However, instead of using the absolute maximum elevation, it subtracts 4000 meters from the maximum. This adjustment is crucial because it addresses a common issue in visualizing topographical data: if there are a few very high points (like mountain peaks), they can skew the color distribution, causing lower but more widespread elevations to lose detail or become visually compressed. This was discussed in more detail in the previous chapter. `norm = BoundaryNorm(bounds, cmap.N)` is used to map these elevation bounds to the color map. This normalization ensures that the color map is applied to the elevation data in a way that reflects the adjusted range, thus avoiding the overshadowing of lower elevations by a few high points.

Figure 10.3: Custom colourmap using the colours of the Pakistani flag.

The rationale behind lowering the top of the normalization range is to enhance the visual distinction of the vast majority of the terrain. By doing this, you ensure that the color map more effectively highlights the variations in the lower and mid-range elevations, which might otherwise be less noticeable if the few extreme elevations were included in the full range. This approach makes for a more informative and visually balanced representation of Pakistan's diverse topography.

```
from matplotlib.colors import ListedColormap
from matplotlib.colors import BoundaryNorm
from matplotlib.colors import LinearSegmentedColormap

cmap = LinearSegmentedColormap.from_list('pakistan', ['black', 'green', '#ffffff'])
newcolors = cmap(np.linspace(0, 1, int((abs(np.amin(clipped_array) +
                                       abs(np.amax(clipped_array)))) / 10)))
newcolors[:1, :] = np.array([1.0, 1.0, 1.0, 1.0])
cmap = ListedColormap(newcolors)

bounds = np.arange(np.amin(clipped_array), np.amax(clipped_array)-4000, 10)
norm = BoundaryNorm(bounds, cmap.N)
```

```
from matplotlib import colors

fig, ax = plt.subplots()
fig.set_size_inches(7, 3.5)

show(clipped_array, ax=ax, transform=clipped_transform, cmap=cmap, norm=norm)
ax.axis('off')
plt.show()
```

Hillshade

Finally we're going to use the *earthpy.spatial* module, specifically the hillshade function, to create a hillshade effect on the clipped raster data representing Pakistan's topography. Hillshading is a technique used in map making and data visualization to simulate shadows and highlights as they would appear on physical terrain, adding a sense of depth and dimensionality to the image. Let's explore the components of this function and its parameters:

- The hillshade function is part of the EarthPy library, which is a set of tools for working with spatial data, particularly raster

Figure 10.4: Topographical map of Pakistan plotted using our custom colourmap.

datasets. The function creates a grayscale image where the brightness of each pixel is determined by its illumination from a hypothetical light source. This results in a shaded relief map that mimics the appearance of sunlight casting shadows over the terrain.

- `hillshade(clipped_array[0], ...)` - Here, you are using the first band ([0]) of your clipped_array, which represents the elevation data of Pakistan after being processed (merged and clipped). This data forms the basis of the hillshade computation.
- *azimuth=240* - This parameter specifies the horizontal angle (in degrees) of the light source relative to the north. An azimuth of 240 degrees means the light is considered to be coming from

the southwest. This direction affects how the shadows and highlights are cast across the terrain features, influencing the overall appearance of relief on the map. The choice of azimuth angle can dramatically change the visual effect and should be chosen based on the desired aesthetic and the geographical context of the data.

- *altitude=1* - This parameter sets the vertical angle or elevation (in degrees) of the light source above the horizon. An altitude of 1 degree is very low, implying that the light source (like the sun) is near the horizon, creating long shadows and a dramatic contrast. This can emphasize the relief and texture of the terrain but may also exaggerate certain features. Adjusting the altitude can help in achieving a balance between drama and realism in the visualization.

The resulting hillshade image is a grayscale representation where the brightness of each pixel indicates the angle at which it is illuminated by the light source, simulating a 3D effect on the 2D raster data. This technique is widely used in geographical information systems (GIS) and mapping to enhance the perception of terrain features, making them more visually accessible and understandable. In the context of your project, applying hillshade to the topography of Pakistan will accentuate its mountains, valleys, and other physical features, providing a more dynamic and engaging visual presentation of the landscape.

```
import earthpy.spatial as es

hillshade = es.hillshade(clipped_array[0], azimuth=240, altitude=1)
```

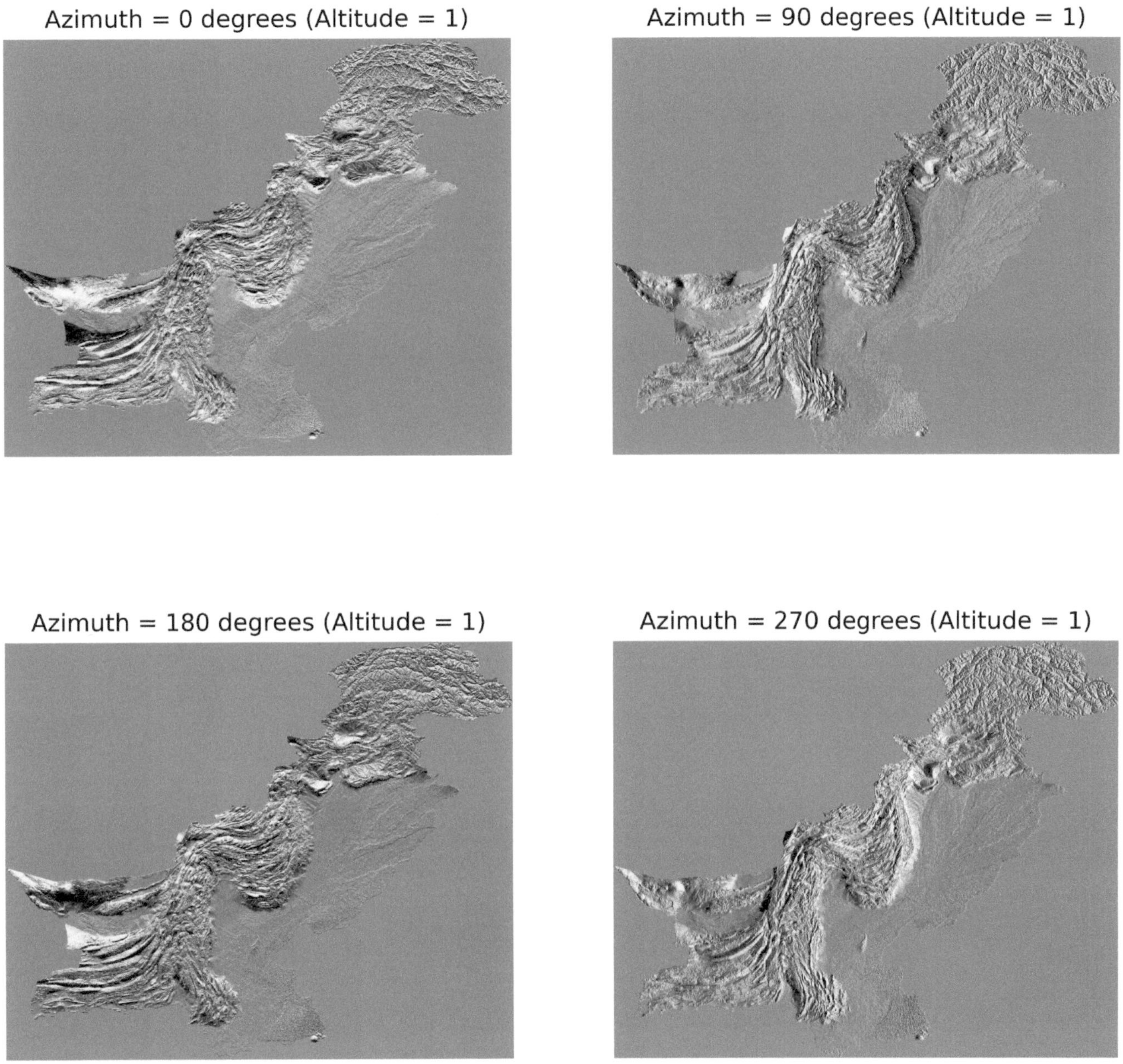

Figure 10.5: Examples of how different Azimuth values affect hillshade.

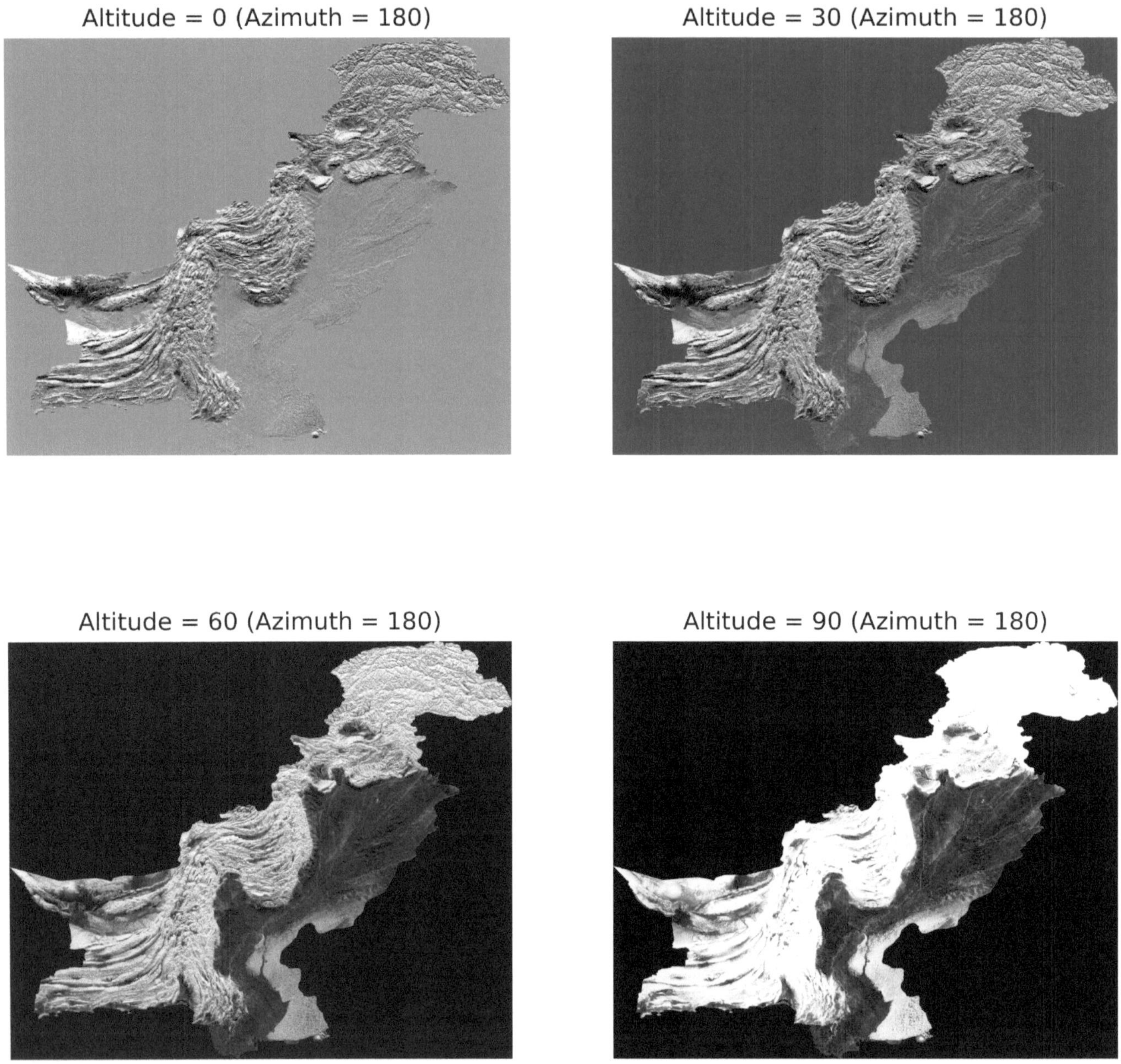

Figure 10.6: Examples of how different Altitude values affect hillshade.

Figure 10.7: Topographical map of Pakistan, using hillshade.

11

Mapping Forests

11.1 Reprojection - Forest Cover

In this chapter, we embark on a journey through the advanced realms of raster operations, specifically, the techniques of reprojecting. Our exploration is set against the backdrop of forest cover analysis, a field where accurate and precise mapping is not just beneficial but essential.

Reprojection is the process of transforming raster data from its original coordinate reference system (CRS) to a different CRS. This transformation is crucial when dealing with spatial data from various sources, each possibly in a different CRS, ensuring consistency and accuracy in spatial analyses. Using Python's rasterio library, we'll walk through the process of reprojecting a forest cover raster. We'll do this first on a global scale and look at projections encountered in previous chapters, e.g. Robinson and then use the lessons learned in the previous chapter to mask the data down to just show the forests of South America, which we will then reproject to a more region specific CRS.

We start by reading a raster file and then printing its minimum and maximum values. The key point of interest here is the discrepancy between the number of unique values in the raster data (102) and the range between the minimum and maximum values (254). This discrepancy is due to the nature of the data: it represents the percentage of forest cover, which naturally ranges from 0 to 100% (accounting for 101 unique values). The additional value, 254, is used to represent 'no data' areas within the raster. This practice is common in raster datasets where specific values are designated to represent the absence of data. In this case, the value 254 allows for a clear distinction between areas with no forest cover (0%) and areas where data is simply not available or applicable.

```
import rasterio
import numpy as np

forests_file = rasterio.open('gm_ve_v1.tif')
forests = forests_file.read()

print(np.amin(forests))
print(np.amax(forests))
print(len(np.unique(forests)))
```

```
0
254
102
```

Plotting the data reveals a problem.

```
from rasterio.plot import show
import matplotlib.pyplot as plt


fig, ax = plt.subplots()
fig.set_size_inches(7, 3.5)
show(forests, ax=ax, cmap='Greens')
ax.axis('off')
plt.show()
```

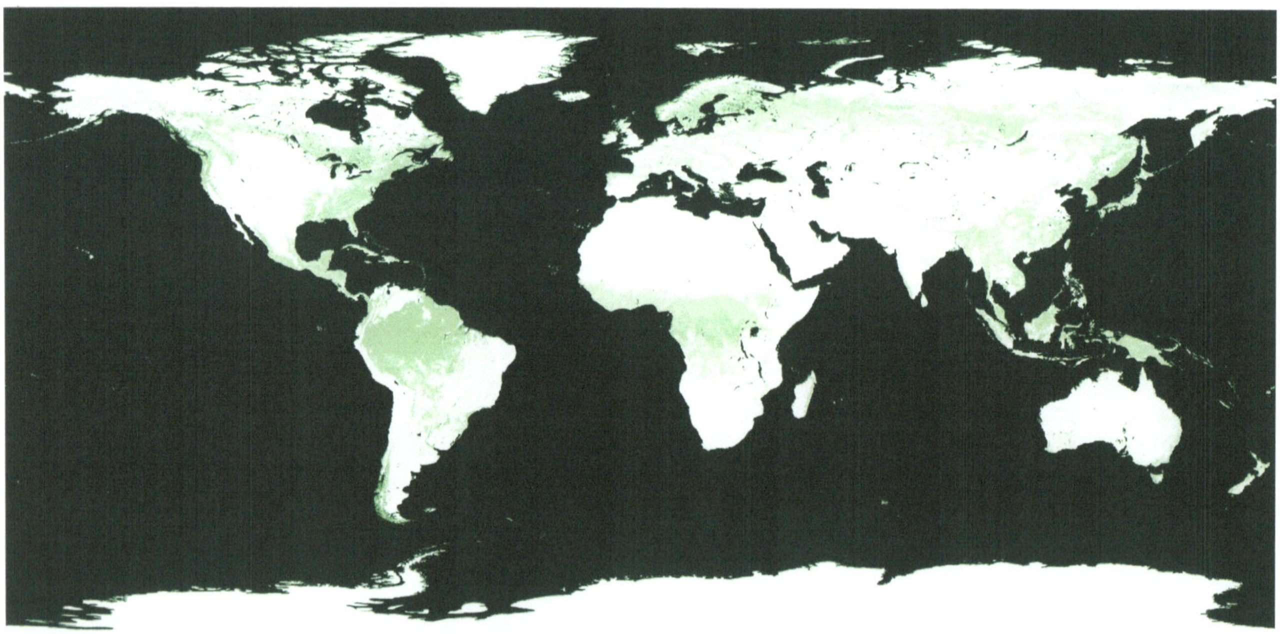

Figure 11.1: Exploratory plot showing our forests data.

The issue we're encountering is a common one when dealing with raster datasets containing both data and designated 'no data' or special value points. In this case, this problem arises due to the way the data is structured and how Matplotlib automatically scales the color map.

The raster dataset includes forest coverage percentages ranging from 0 to 100, and an additional unique value of 254 to represent water bodies. When this is plotted, Matplotlib linearly scales the color map from the lowest value (0) to the highest (254). The linear scaling of the color map means that the actual forest coverage data, which should ideally occupy the full spectrum of the color map, is compressed into roughly the first 40% of the color range. This is because the data of interest (0–100%) is only a portion of the entire range that Matplotlib considers (0–254). Consequently, a significant portion of the color map (60%) is used to represent the value 254, assigned to water bodies. This scaling issue leads to a visual misrepresentation where the oceans, marked as 254 in the dataset, are displayed in the color that would typically represent the highest forest coverage. Essentially, areas without forest cover (oceans in this case) are misleadingly depicted with a green hue, suggesting forest presence where there is none. Moreover, there is a large range of colors (those representing values 101–253) that are completely unused. This is because there are no actual data points within this range in your dataset. As a result, a substantial portion of the color map is effectively wasted, further skewing the visual accuracy of the map. To address this, we can modify the color map and normalization to accurately reflect the data's range.

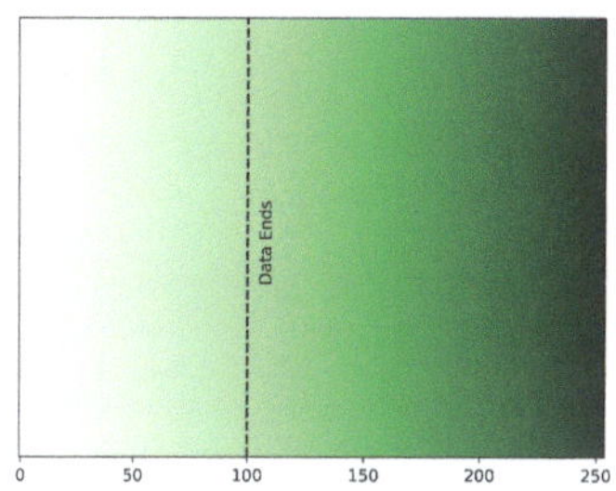

Figure 11.2: Schematic showing the relationship between our colourmap and our data.

The line `forests[forests == 254] = 0` is used to replace the 'no data' values, originally marked as 254, with 0. This adjustment simplifies the dataset by ensuring that non-forest areas (such as oceans) and 'no data' areas are represented by the same value, making the subsequent color mapping process more straightforward.

```
forests[forests == 254] = 0
```

`bounds = np.arange(102)` defines the boundaries for the normalization, covering the range of 0 to 101, which now includes the 'no data' value at 0. We retrieve the *Greens* color map limited to 101 colors to represent the 0–100% range of forest cover. `newcolors = greens( np.linspace(0, 1, len(bounds)))` creates an array of colors, linearly spaced within the 'Greens' color map between 0 and 101 (the values in bounds). The first color (representing the lowest value, which is now 0 for both non-forest and 'no data' areas) is set to a transparent color (*background_colour*), allowing you to visually distinguish these areas from actual forest cover. A new ListedColormap is then created from these colors (*newcmp_forests*).

`norm_forests = BoundaryNorm(bounds, newcmp_forests.N)` creates a

`BoundaryNorm` instance, which maps these boundary values to the indices in the custom colormap. This normalization ensures that each percentage of forest cover is mapped to the correct color in the 'Greens' spectrum.

By creating this custom color map and normalization and adjusting the 'no data' values, we ensure that the forest cover visualization is both accurate and visually clear. The 'Greens' spectrum is now used exclusively for depicting actual forest cover percentages, and areas with no data or no forest cover are effectively and clearly differentiated from forested areas.

```
from matplotlib.colors import ListedColormap, LinearSegmentedColormap, BoundaryNorm
from matplotlib import cm

bounds = np.arange(102)
greens = cm.get_cmap('Greens', len(bounds))
newcolors = greens(np.linspace(0, 1, len(bounds)))
background_colour = np.array([0.0, 0.0, 0.0, 0.0])
newcolors[:1, :] = background_colour
newcmp_forests = ListedColormap(newcolors)
norm_forests = BoundaryNorm(bounds, newcmp_forests.N)
```

```
from rasterio.plot import show
fig, ax = plt.subplots()
fig.set_size_inches(7, 3.5)
show(forests, ax=ax, cmap=newcmp_forests, norm=norm_forests)

ax.axis('off')
plt.show()
```

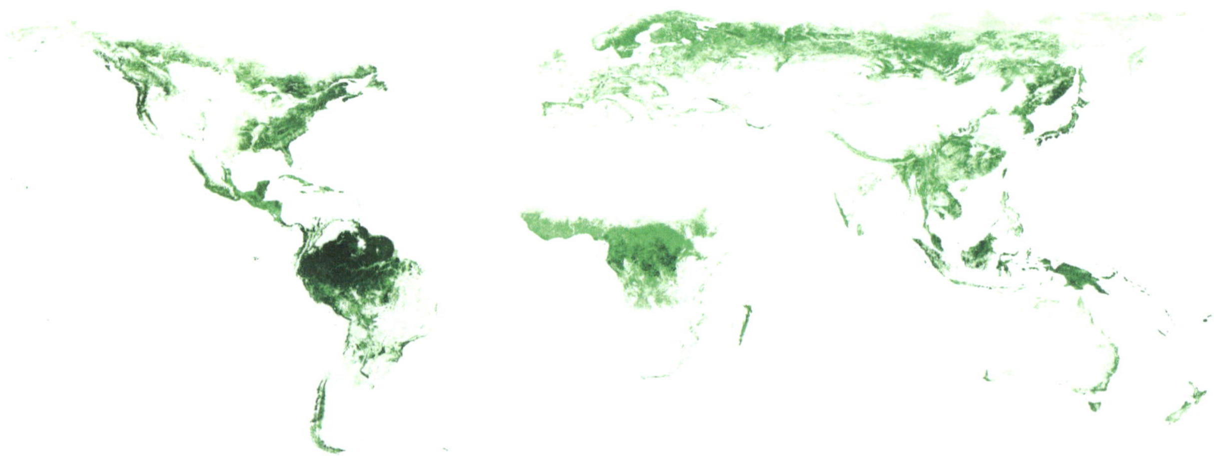

Figure 11.3: Global forests.

Reprojection

The Robinson projection is one of the more realistic ways to plot global datasets, accounting for the fact that the earth is not flat. We are going to show a couple of ways to reproject rasters. The first uses rioxarray and the second uses rasterio.

The code uses the *rioxarray* and rasterio libraries, which are powerful tools for handling and manipulating geospatial data in Python. The rioxarray library extends xarray with the *rio* accessor and allows rasters from tiff and geotiff files to be clipped, merged and reprojected. Below, the original tif file is opened with rioxarray (it leverages some underlying rasterio functionality, hence the `open_rasterio` method), the Robinson projection is defined as a `rasterio.CRS` object and then the raster is reprojected onto this CRS. One thing to note, the `rasterio.CRS` object is created from a pyproj string. pyproj is a fantastic library that deals with cartographic projections and coordinate transformations.

`forests_file = rxr.open_rasterio('gm_ve_v1.tif', masked=True).squeeze()` This line uses `rxr.open_rasterio` to load a raster file (*gm_ve_v1.tif*). The *masked=True* argument indicates that the function should read the data and automatically mask out 'no data' values. The `.squeeze()` method is used to drop any single-dimensional entries from the shape of the returned array, simplifying its structure.

`crs_rob = CRS.from_string('+proj=robin')` defines a new CRS using the `CRS.from_string` method from rasterio.crs. The *+proj=robin* keyword indicates that the Robinson projection is being used, which is a commonly used projection for world maps due to its balance in representing area and shape.

`forests = forests_file.rio.reproject(crs_rob)` reprojects the forest cover data from its original CRS to the Robinson projection. The `rio.reproject` method is a part of the rioxarray integration with rasterio and is used for performing such reprojections. This step is crucial when the goal is to analyze or display the data in a different spatial context or when combining it with other data in a common CRS. Finally, `forests = forests.to_numpy()` converts the reprojected raster data into a NumPy array. This conversion is often done to leverage the broad functionalities of NumPy for further data manipulation and analysis, or for compatibility with other Python libraries and functions that require input data in array format.

```
import rioxarray as rxr
```

```
from rasterio.crs import CRS

forests_file = rxr.open_rasterio('gm_ve_v1.tif', masked=True).squeeze()
print("Original projection: ", forests_file.rio.crs)
crs_rob = CRS.from_string('+proj=robin')
forests = forests_file.rio.reproject(crs_rob)
print("New projection: ", forests.rio.crs)
forests = forests.to_numpy()
```

```
Original projection:  EPSG:4326
New projection:  PROJCS["unknown",
GEOGCS["unknown",
  DATUM["WGS_1984",
   SPHEROID["WGS 84",6378137,298.257223563,AUTHORITY["EPSG","7030"]],
   AUTHORITY["EPSG","6326"]
  ],
  PRIMEM["Greenwich",0,AUTHORITY["EPSG","8901"]],
  UNIT["degree",0.0174532925199433,AUTHORITY["EPSG","9122"]]
],
  PROJECTION["Robinson"],
  PARAMETER["longitude_of_center",0],
  PARAMETER["false_easting",0],
  PARAMETER["false_northing",0],
  UNIT["metre",1,AUTHORITY["EPSG","9001"]],
  AXIS["Easting",EAST],
  AXIS["Northing",NORTH]
]
```

```
forests[forests > 100] = 0.0

fig, ax = plt.subplots()
fig.set_size_inches(7, 3.5)

show(forests, ax=ax, cmap=newcmp_forests, norm=norm_forests)
ax.axis('off')

plt.show()
```

Masking and Reprojection

We are going to clip our forests file to just that of South America and then reproject that raster using rasterio.

As in previous chapters, we need to load our land polygons and in this case we are indexing to those of South America.

```
import geopandas as gpd
import pandas as pd

df = gpd.read_file('NaturalEarth/10m_cultural/ne_10m_admin_0_countries.shp')

samerica = df.loc[df['CONTINENT'] == 'South America']
```

As in previous chapters, we are going to use the rasterio mask function to clip our raster to only the data within our South America polygons. Here we are setting nodata to 0 as we already know that areas of no forests should be defined as 0% forest.

```
import rasterio
```

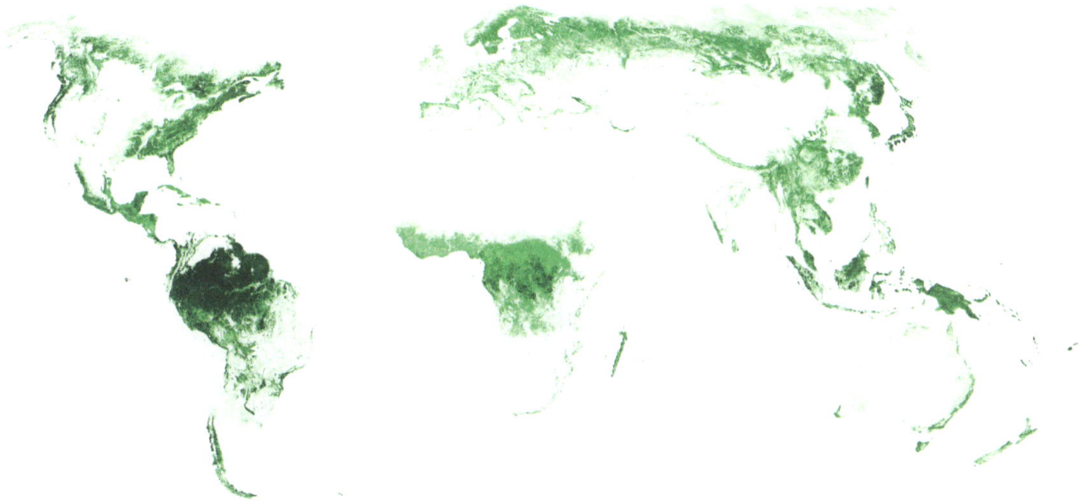

Figure 11.4: Global forests reprojected to the Robinson projection.

```
from rasterio import mask as msk
from shapely.geometry import mapping


clipped_array, clipped_transform = rasterio.mask.mask(forests_file, [
    mapping(geom) for geom in (samerica.geometry.tolist())
], crop=True, nodata=0)
```

This looks similar to previous raster plots but we are adding a new element by plotting the outlines of the countries of South America.

The `samerica.boundary.plot(ax=ax, edgecolor='black', lw=0.4)` command is used to overlay the boundary of the specified region on top of the raster plot. This is achieved using the *.boundary* attribute and the `.plot` method from GeoPandas. The *edgecolor='black'* parameter outlines the boundary in black, and *lw=0.4* sets the line width, ensuring the boundary is clearly visible but not overpowering. Plotting the boundary in this manner provides a clear delineation of the region of interest, enhancing the geographical context and interpretability of the raster data.

```
clipped_array[clipped_array >= 101] = 0.0

fig, ax = plt.subplots()
fig.set_size_inches(7, 3.5)

show(clipped_array[0], ax=ax, transform=img_meta['transform'],
     cmap=newcmp_forests, norm=norm_forests)
```

```
samerica.boundary.plot(ax=ax, edgecolor='black',  lw=0.4)

ax.axis('off')
plt.show()
```

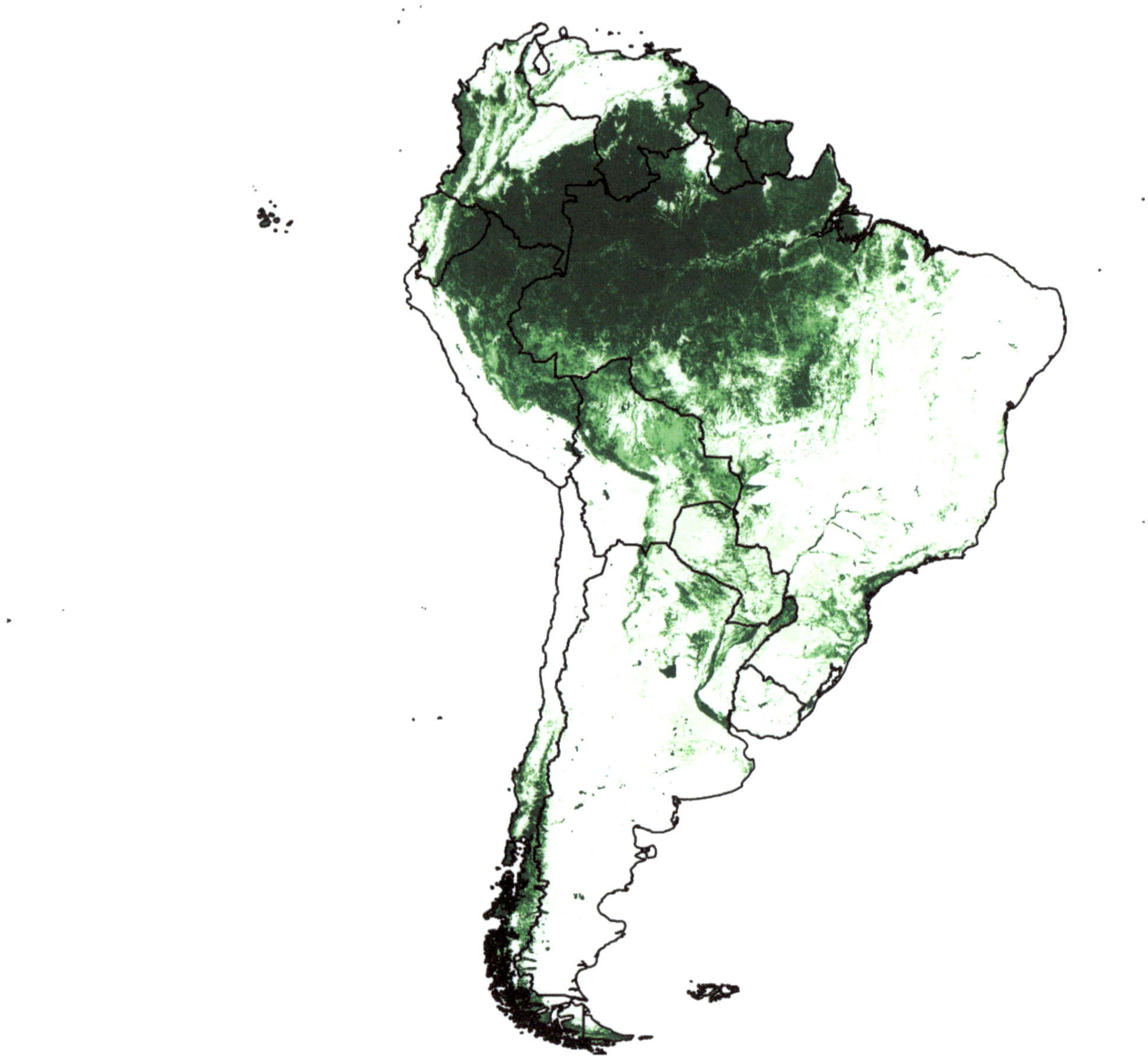

Figure 11.5: South American forests map with country boundaries overlaid.

Lets also put a colourbar to add additional context. Here, we are adding a colorbar to the plot to represent the range of tree cover percentages in the raster data. A `ScalarMappable` object is created with the custom forest color map (*newcmp_forests*) and normalization (*norm_forests*), which is then used to generate the colorbar. The colorbar is configured to be horizontal and placed within the plot area using specific axes dimensions. It is formatted to show tick marks at key percentages (0%, 25%, 50%, 75%, 100%) for clear interpretation of the color gradations in terms of tree cover percentage. The colorbar is further customized with label sizes, colors, and an additional label indicating that it represents the 'Percentage Tree Cover.' This addition enhances the readability and interpretability of the plot, providing viewers with a clear understanding of what the colors represent in terms of forest density.

```
fig, ax = plt.subplots()
fig.set_size_inches(7, 3.5)

show(clipped_array[0], ax=ax, transform=img_meta['transform'],
     cmap=newcmp_forests, norm=norm_forests)
samerica.boundary.plot(ax=ax, edgecolor='black',  lw=0.4)

fig = ax.get_figure()
cax = fig.add_axes([0.55, 0.2, 0.1, 0.01])
sm = plt.cm.ScalarMappable(cmap=newcmp_forests, norm=norm_forests)
sm._A = []
cb = fig.colorbar(sm, cax=cax, orientation="horizontal", pad=0, format='%.3g',
                  ticks=[1, 25, 50, 75, 100], drawedges=False)
cb.ax.set_xticklabels(['0%', '25%', '50%', '75%', '100%'])
cb.outline.set_visible(False)
cb.ax.tick_params(labelsize=4, width=0.2, length=0.75, color='black')
cbytick_obj = plt.getp(cb.ax, 'xticklabels' ) #Set y tick label color
plt.setp(cbytick_obj, color='black')
cb.ax.set_xlabel('Percentage Tree Cover', fontsize=4, color='black', labelpad=-20)
ax.axis('off')

plt.show()
```

As we move into the next section of our exploration in advanced raster operations, we encounter the common challenge of managing data formats while performing complex tasks in Python. Specifically, we will be focusing on the reprojection of a clipped raster. This process often involves navigating between different data structures: from numpy arrays, which are efficient for manipulation and computation, to rasterio objects, which are essential for spatial operations like reprojection. To bridge this gap, our initial step will involve creating a memory file to house our clipped raster data, currently in a numpy array format. This memory file acts as an intermediary, enabling us to effectively utilize rasterio's capabilities for reprojection.

```
from rasterio.io import MemoryFile

memfile = MemoryFile()
dataset = memfile.open(driver='GTiff', height=img_meta['height'],
                       width=img_meta['width'],
                       count=1, crs=forests_file.crs,
                       transform=clipped_transform, dtype=clipped_array.dtype)
merged_data = dataset.write(clipped_array[0], 1)
```

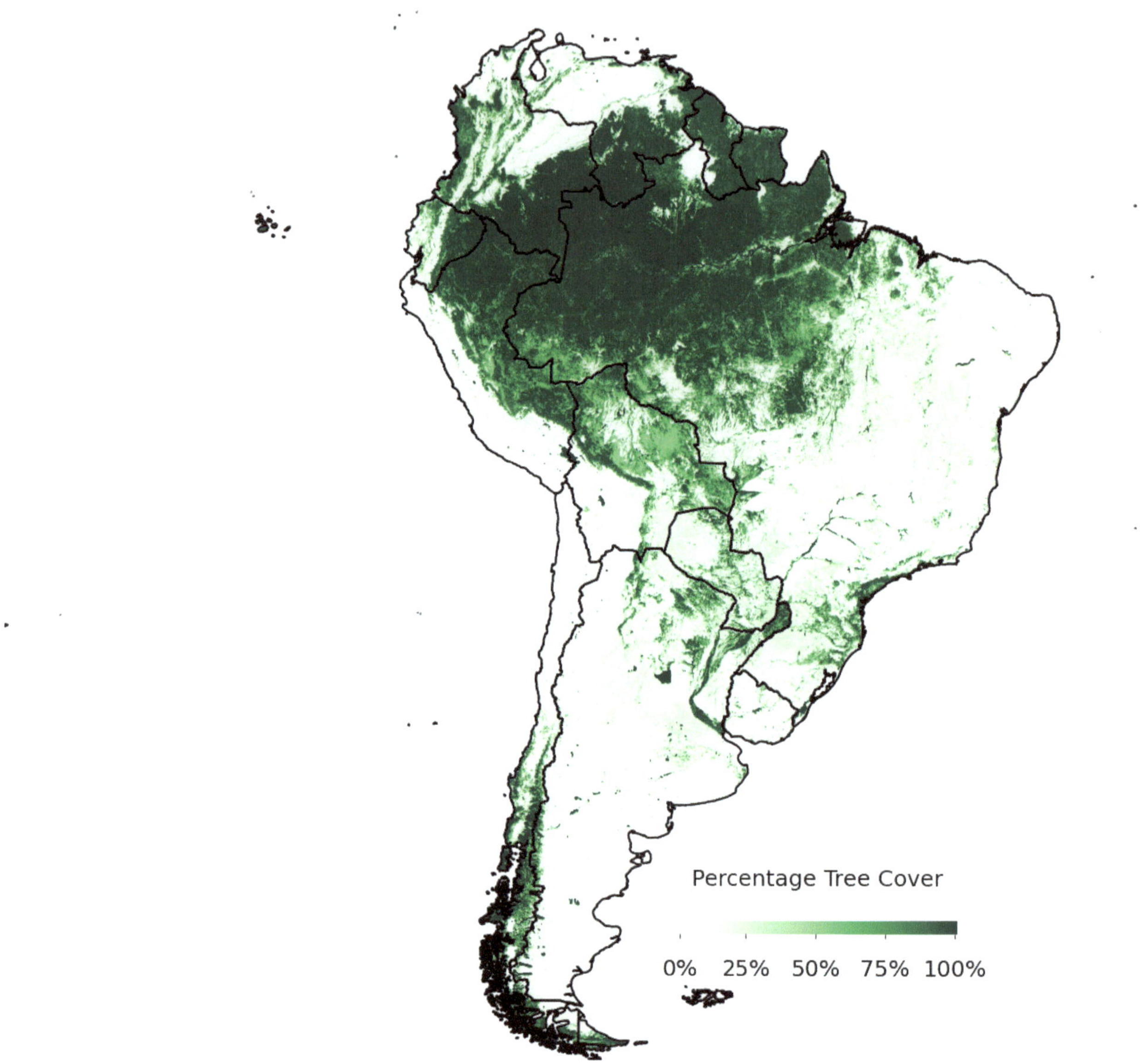

Figure 11.6: South American forests map with country boundaries overlaid and a colourmap.

In this section, we're demonstrating a different method of raster reprojection, distinct from the previous use of rioxarray. Here, we utilize core functionalities from the rasterio library, a robust tool for handling raster data in Python. This alternative approach showcases another efficient method for transforming the spatial reference of a raster dataset.

The code imports Resampling and reproject from rasterio, along with a function to calculate the default transformation for the new projec-

tion. `dst_crs = 'EPSG:3395'` sets the destination coordinate reference system (CRS) to
EPSG:3395, which is the Mercator projection. This CRS is widely used for mapping purposes due to its ability to represent direction consistently and is a good choice for many mapping applications.
`calculate_default_transform` is used to compute the transformation parameters and new dimensions needed to reproject the dataset into the Mercator projection. This function takes the original CRS of the dataset, the destination CRS, original dimensions, and bounds of the dataset to calculate the necessary transformation.

An empty numpy array dst_array is created with the new height and width calculated for the Mercator projection. This array will hold the reprojected raster data. The data type (dtype) of the array matches that of the clipped_array to ensure consistency.

The reproject function from rasterio.warp is used to execute the reprojection. Key parameters include:

- *source*: Specifies the input raster data. Here, rasterio.band(dataset, 1) is used to select the first band from the dataset.
- *destination*: The numpy array (*dst_array*) where the reprojected data will be stored.
- *src_transform* and *src_crs*: The original transform and CRS of the dataset.
- *dst_transform* and *dst_crs*: The new transform and CRS for the reprojected data.
- *resampling*: The resampling method, set to `Resampling.nearest`. This parameter determines how pixel values are interpolated when changing the raster's resolution or projection. Nearest neighbor resampling is a common choice for categorical or discrete data.

```
from rasterio.enums import Resampling
from rasterio.warp import calculate_default_transform, reproject

# Define the destination CRS - Mercator projection
dst_crs = 'EPSG:3395'

# Calculate the transform and dimensions for the new projection
transform, width, height = calculate_default_transform(dataset.crs, dst_crs,
                                                      dataset.width,
                                                      dataset.height,
                                                      *dataset.bounds)

# Create a destination numpy array
dst_array = np.empty((height, width), dtype=clipped_array.dtype)

```

```
# Perform the reprojection
reproject(
    source=rasterio.band(dataset, 1),
    destination=dst_array,
    src_transform=dataset.transform,
    src_crs=dataset.crs,
    dst_transform=transform,
    dst_crs=dst_crs,
    resampling=Resampling.nearest
)
```

```
dst_array[dst_array >= 101] = 0.0

fig, ax = plt.subplots()
fig.set_size_inches(7, 3.5)

show(dst_array, ax=ax, transform=transform, cmap=newcmp_forests, norm=norm_forests)

ax.axis('off')

plt.show()
```

Plotting the data alongside the previous raster shows the effect of different projections.

```
# Plot raster
fig, (ax1, ax2) = plt.subplots(nrows=1, ncols=2, figsize=(10, 4))

show(clipped_array[0], ax=ax1, transform=img_meta['transform'],
     cmap=newcmp_forests, norm=norm_forests)
ax1.axis('off')

show(dst_array, ax=ax2, transform=transform, cmap=newcmp_forests, norm=norm_forests)
ax2.axis('off')

plt.tight_layout()
plt.show()
```

Figure 11.7: South American forests re-projected to the Mercator projection.

Figure 11.8: Side by side view of the two projections.

12

Mapping River and River Basins

12.1 Rivers Basin Maps

This chapter delves into an exciting realm where we combine the precision of vector data with the depth of raster imagery to create intricate river maps. Building on the foundations laid out in previous chapters on plotting rasters and vector data.

In this chapter, we will treat rivers with the reverence they deserve. We will learn to plot them not just as lines, but as dynamic entities with varying sizes, each colored uniquely to represent their individual basins. These visual cues will not only enhance the aesthetic appeal of our maps but also provide deeper insights into the geographical narratives of these watercourses.

Our journey will start with a deep dive into vector data manipulation using GeoPandas. We will explore advanced techniques to scale river representations based on their size, providing a more accurate portrayal of their real-world significance. We'll then move on to applying a spectrum of hues to differentiate river basins, thereby adding a layer of information that is as informative as it is visually engaging.

The final section of this chapter is the integration of these enhanced vector representations onto a topographic map. Here, raster data comes into play, offering a rich, textured backdrop of hills, valleys, and plains. By plotting our meticulously crafted river data onto this canvas, we reveal the relationship between waterways and the landscapes they wind through. This fusion of data types does more than just map rivers; it tells a story of how these watercourses shape, and are shaped by, the topography they meander across.

As we navigate through this chapter, we will arm ourselves with advanced Python coding techniques, a keen eye for aesthetic detail, and a deep appreciation for the geospatial storytelling power of maps. By the end, you will not only be adept at creating visually stunning and information-rich river maps but also be inspired by the potential of Python in transforming raw data into compelling geographic narratives.

We are first loading the data using GeoPandas. The data comes from the Hydrosheds project and can be downloaded as shapefiles corresponding to certain continents.

```
import geopandas as gpd

africa_rivers = gpd.read_file("a00000009.gdbtable")

print(africa_rivers.head())
```

As in previous chapters, we need to load our land polygons and in this case we are indexing to those of Africa.

```
country_shapefiles = gpd.read_file(
    "../PythonMapsData/NaturalEarth/10m_cultural/ne_10m_admin_0_countries.shp")
africa = country_shapefiles.loc[country_shapefiles['CONTINENT'] == 'Africa']
```

In the pursuit of creating accurate and visually coherent river maps, we sometimes encounter challenges that necessitate clever data manipulation techniques. One such challenge arises from the way rivers interact with lakes in our datasets. Ideally, lakes should appear as distinct, empty regions on our map, clearly separated from river paths. However, due to the way this data was generated, we have lines which intersect in our lakes, spoiling the visual.

To overcome this, we will incorporate the lakes dataset from NaturalEarth. We load this dataset as a `GeoDataFrame` of polygons, each representing an individual lake. However, to manage data volume and maintain focus, we limit our scope to the lakes in Africa.

We perform a spatial join between the lakes dataset and the African polygons loaded earlier. This operation utilizes GeoPandas's join method, merging the datasets based on their spatial relationship—in this case, the intersection of lake polygons with the African continent.

Having isolated the African lakes, we now turn our attention to the river dataset. The goal is to identify and remove those river segments that erroneously traverse lakes, resulting from the limitations of our topography data. This is achieved through another spatial join, this time between our African rivers dataset and the African lakes polygons.

The within operation in `sjoin` allows us to capture those river segments that fall within the lake polygons, signifying areas of potential data distortion.

The final step involves purging these inaccurately represented river segments from our primary river dataset. By excluding the indices of the identified segments, we refine our river dataset, ensuring that it accurately represents the true paths of rivers, free from the distortions caused by intersecting lake data.

```
lakes = gpd.read_file("../PythonMapsData/NaturalEarth/10m_physical/ne_10m_lakes.shp")
lakes = lakes.loc[lakes['featurecla'] == 'Lake']
lakes = lakes.dropna(subset=['name'])

africa_lakes = gpd.sjoin(lakes, africa, op='intersects')
africa_lakes = lakes.loc[lakes.index.isin(africa_lakes.index.tolist())]
excluded_rivers = gpd.sjoin(africa_rivers, africa_lakes, op='within')
africa_rivers = africa_rivers.loc[~africa_rivers.index.isin(excluded_rivers.index.tolist())]
```

We are also going to exclude rivers with an *ORD_FLOW* value greater than 8. These values essentially correspond to streams that spend most, if not all of the year dried up.

```
africa_rivers = africa_rivers.loc[africa_rivers['ORD_FLOW'] < 8]
```

The data can now be plotted to give us a sense of what it looks like. I have plotted the lakes as well just for completeness. The data looks vaguely as expected with a massive void where the Sahara should be. With that said, the Kalahari desert is looking disappointingly moist, this will be addressed in due course. Note the voids where lakes should be, notably Lake Victoria in the central, eastern region.

```
import matplotlib.pyplot as plt

fig, ax = plt.subplots()
fig.set_size_inches(5, 7)

africa_rivers.plot(ax=ax, color='blue', lw=0.1)
africa_lakes.plot(ax=ax, color='white', alpha=1)

ax.axis('off')
plt.show()
```

In this segment of the chapter on river mapping, we're focusing on scaling the size of rivers based on a specific attribute—'DIST_UP_KM', which represents the distance upstream. This scaling is crucial for visualizing rivers in a way that reflects their real-world significance. Let's break down what each part of the code does:

`leftSpan = np.amax(africa_rivers['DIST_UP_KM']) - np.amin(africa_rivers['DIST_UP_KM'])` calculates the range of the 'DIST_UP_KM' values in the africa_rivers `DataFrame`. It subtracts the

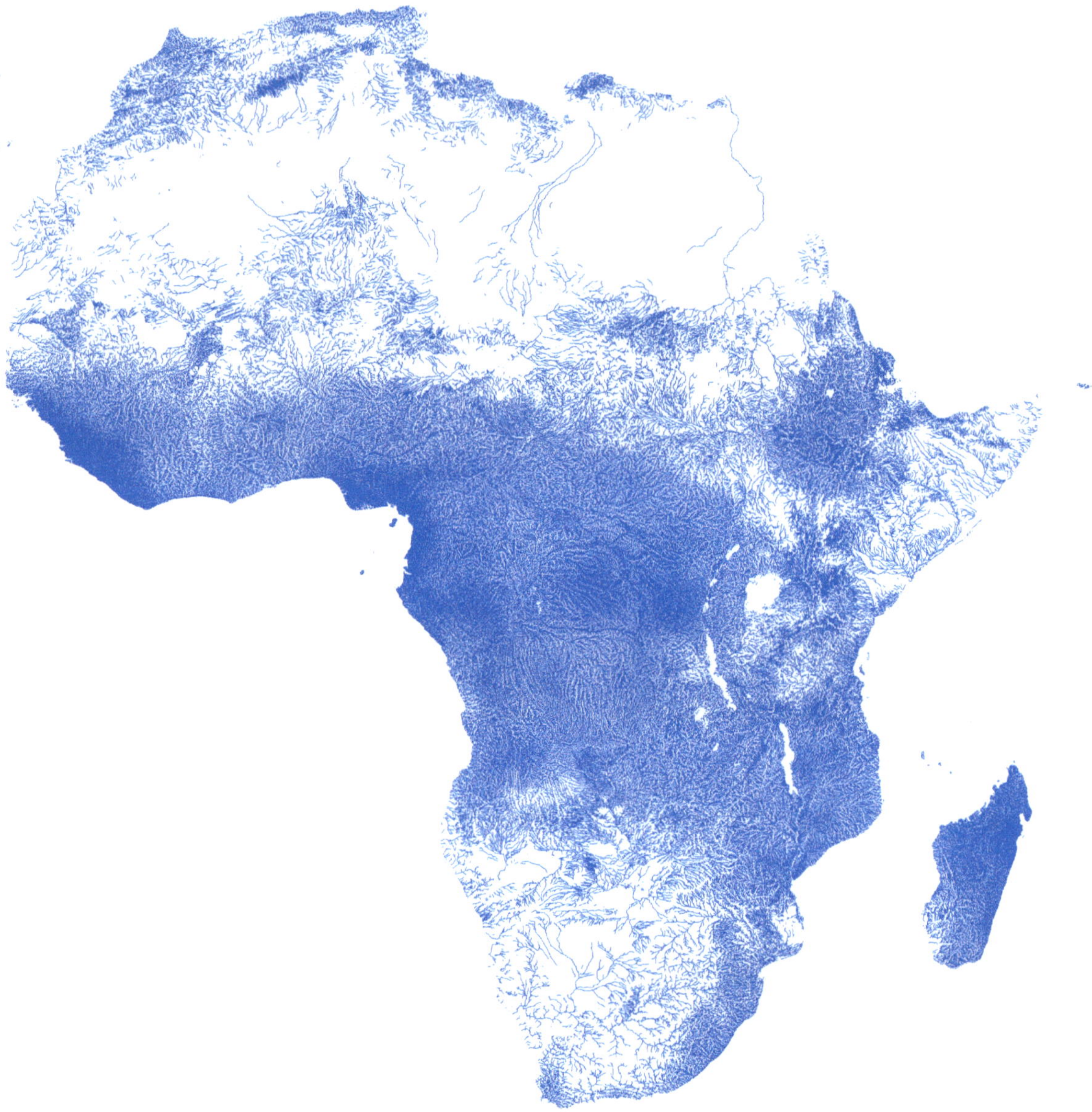

Figure 12.1: African Rivers.

minimum value (*np.amin*) from the maximum value (*np.amax*) of the 'DIST_UP_KM' column. This range (referred to as *leftSpan*) represents the span of the distance upstream across all rivers in the dataset. `rightSpan = 0.6 - 0.04` defines the range of the scaled values. In this case, we want the line widths (representing river sizes) to vary between 0.04 and 0.6. This range (referred to as *rightSpan*) will be used to scale the

'DIST_UP_KM' values.

`valueScaled = (africa_rivers['DIST_UP_KM'] - np.amin(africa_rivers['DIST_UP_KM'])) / leftSpan` normalises the 'DIST_UP_KM' values by subtracting the minimum value and then dividing by the range (*leftSpan*). This normalizes the values between 0 and 1, creating a proportional scale.

`africa_rivers['LW_DIST_UP_KM'] = 0.04 + (valueScaled * rightSpan)` maps these normalized values to our desired line width range. The normalized values are multiplied by *rightSpan* and then offset by the minimum line width (0.04). The resulting values, which now fall within our specified line width range, are stored in a new column 'LW_DIST_UP_KM' in the africa_rivers `DataFrame`. By executing these steps, each river in the *africa_rivers* `DataFrame` is assigned a line width that corresponds to its 'DIST_UP_KM' value. The rivers with greater 'DIST_UP_KM' values (indicating a longer distance upstream) will be represented by thicker lines, while those with smaller values will have thinner lines. This visual representation allows for an intuitive understanding of the rivers' relative sizes and importance within the map, enhancing the overall interpretability and aesthetic appeal of the river map.

```
import numpy as np

leftSpan = np.amax(africa_rivers['DIST_UP_KM']) - np.amin(africa_rivers['DIST_UP_KM'])
rightSpan = 0.6 - 0.005
valueScaled = (africa_rivers['DIST_UP_KM'] -
               np.amin(africa_rivers['DIST_UP_KM'])) / leftSpan
africa_rivers['LW_DIST_UP_KM'] = 0.005 + (valueScaled * rightSpan)
```

We can then visualise Africa rivers dataset on a map, with the line widths of the rivers dynamically adjusted based on their respective 'DIST_UP_KM' values, as processed earlier. The code uses GeoPandas' plotting capabilities to achieve this visualization:

`africa_rivers.plot(ax=ax, ...)` is the core of the plotting operation. It tells Python to plot the *africa_rivers* `GeoDataFrame` on the provided Matplotlib axes (ax). The `GeoDataFrame` *africa_rivers* contains the geographic data for the rivers. *edgecolor='face'* sets the edge color of the lines (representing the rivers) to match the face color. In simpler terms, it removes the border color of the lines, ensuring a more uniform appearance of the rivers on the map. *color='blue'* sets the color of the lines to blue, a conventional choice for depicting water bodies like rivers. This ensures that the rivers are easily distinguishable and visually accurate on the map. Crucially, `lw=africa_rivers['LW_DIST_UP_KM']` sets the line width (lw) for each river. Instead of a constant value, the line width for each river is determined by its corresponding value in the 'LW_DIST_UP_KM' column of the africa_rivers DataFrame. This column contains scaled values based on the 'DIST_UP_KM' attribute, as calculated in the previous step. As a result, the width of each river line on the map will reflect its relative size or significance, with larger rivers (having greater 'DIST_UP_KM' values) being represented by thicker lines.

```
fig, ax = plt.subplots()
fig.set_size_inches(3.5, 7)

africa_rivers.plot(ax=ax, edgecolor='face', color='blue', lw=africa_rivers['LW_DIST_UP_KM'])
africa_lakes.plot(ax=ax, color='white')

ax.axis('off')
plt.show()
```

River Basins

In this phase, we load and briefly inspect the river basins dataset for Africa to understand its structure and content. The dataset includes key columns such as "MAJ_NAME" and "SUB_NAME", which are essential for our analysis. The "MAJ_NAME" column represents the major river basins, grouping together all rivers and tributaries that contribute to a major river system, such as the Nile. On the other hand, "SUB_NAME" provides information on smaller, sub-basins within these major basins, offering a more detailed view of the river network's segmentation. Understanding the distinction between these two columns is crucial for our analysis, as it allows us to visualize and analyze the river networks at different hierarchical levels, from major basins down to their smaller constituent sub-basins.

```
import geopandas as gpd

basins = gpd.read_file("../Nature/rivers/data/basins/africa/hydrobasins_africa.shp")
print(basins.head())
```

```
SUB_BAS  TO_BAS  MAJ_BAS  SUB_NAME  MAJ_NAME  SUB_AREA  MAJ_AREA
0    90002    -888     7009         Borkou    Lake Chad   1179610   2461890
1    30003   30272     7003   White Nile 5         Nile     97222   3074955
2    90008    -888     7009         Dillia    Lake Chad    162385   2461890
3    30029   30297     7003  Lake Victoria         Nile     69200   3074955
4   100030    -888     7010     Lake Rukwa  Rift Valley      3443    638878

LEGEND                                           geometry
0       9  POLYGON ((15.52500 13.52083, 15.52917 13.52083...
1       3  MULTIPOLYGON (((32.24167 6.16250, 32.25000 6.1...
2       9  POLYGON ((12.02917 16.12083, 12.03333 16.12083...
3       3  POLYGON ((33.26667 0.37500, 33.26667 0.37083, ...
4      10  POLYGON ((31.95000 -7.55000, 31.95000 -7.55833...
```

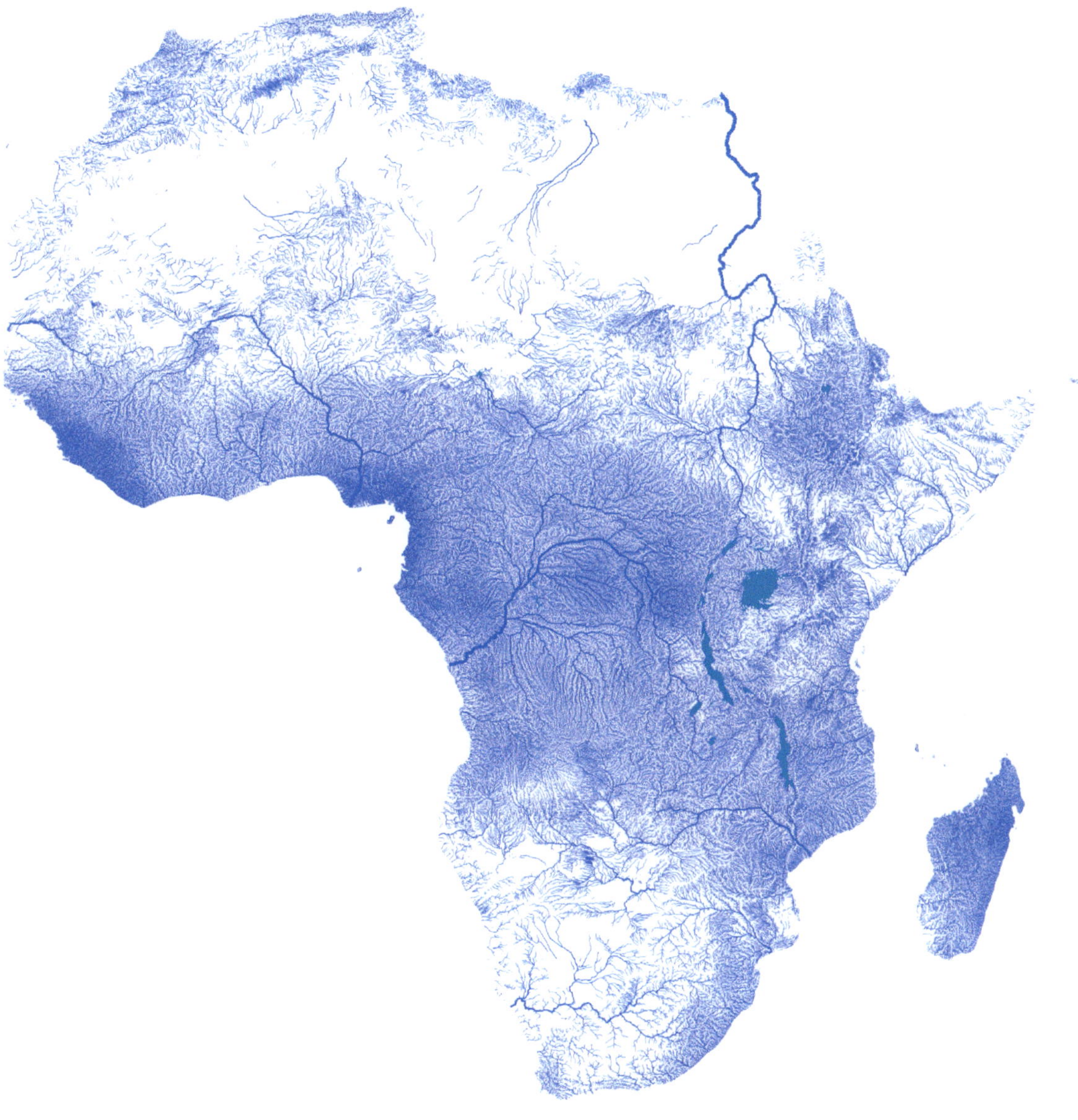

Figure 12.2: African rivers scaled according to their size.

This data can be plotted to make it easier to understand. Here we are making two side-by-side maps to visually differentiate between the smaller sub-basins (SUB_NAME) and the major basins (MAJ_NAME) of Africa's river systems. The first subplot (ax1) is dedicated to plotting the SUB_NAME data, coloring each sub-basin uniquely to highlight the detailed segmentation within the larger river systems. The second subplot (ax2) focuses on the MAJ_NAME data, where each major basin is colored differently, showcasing the broader, overarching river networks.

```
import matplotlib.pyplot as plt

fig = plt.figure()
ax1 = plt.subplot(1, 2, 1)
basins.plot(ax=ax1, column='SUB_NAME', edgecolor='face')
ax1.axis('off')

ax2 = plt.subplot(1, 2, 2)
basins.plot(ax=ax2, column='MAJ_NAME', edgecolor='face')
ax2.axis('off')

plt.show()
```

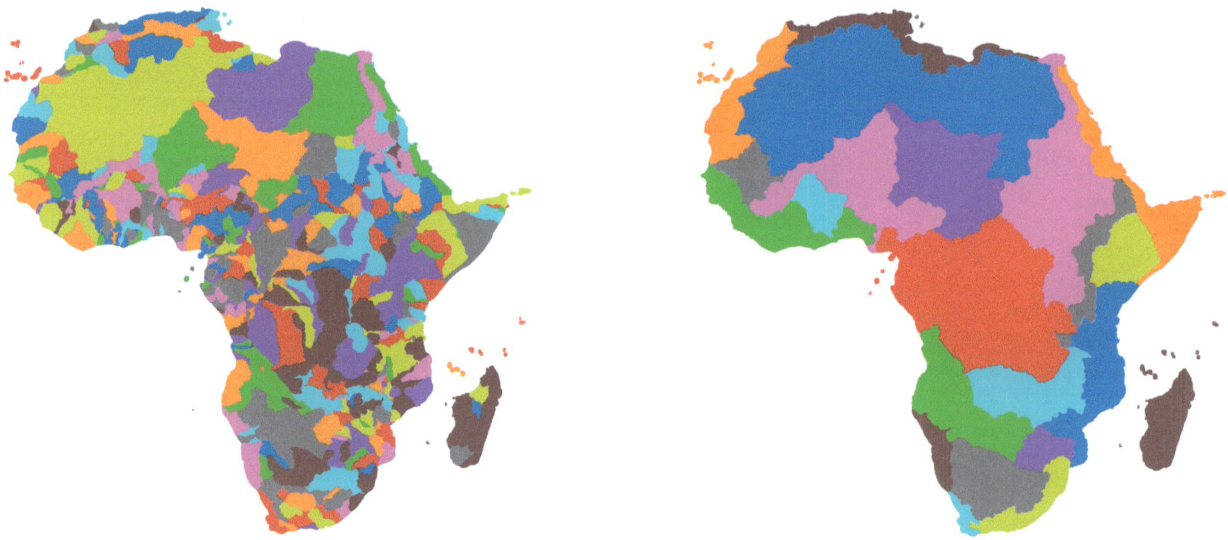

Figure 12.3: African river basins.

Next, we will refine the visualization of the major river basins (MAJ_NAME) by applying a custom color palette to enhance the aesthetic appeal and clarity of the plot. The decision to focus on the major basins is driven by the need for a more coherent and less cluttered visual representation, given the overwhelming number of smaller basins. There are 1450 unique basins in the dataset, which is far too many to plot and colour differently. There are however only 26 major basins e.g. the Nile basin, which contains lots of smaller basins that all house rivers that eventually feed into the Nile.

We start by defining a list of preferred colors in *colours_i_like*. This list includes a variety of colors specified in different formats (like named colors, hexadecimal codes, etc.), providing a rich and diverse palette. Next, we create a new DataFrame, *colors_df*, which maps each unique

MAJ_NAME from the basins `DataFrame` to a color from your custom list. This is achieved by pairing the unique major basin names with the colors.

Finally, we merge this new *colors_df* `DataFrame` with the original basins DataFrame. The merge is performed on the MAJ_NAME column, effectively associating each major basin with a specific color from your preferred palette. This step is crucial as it adds a new column to the basins DataFrame, where each row now includes the color corresponding to its major basin. This enhancement not only personalizes the visualization to your taste but also aids in distinguishing between different major basins more effectively in the subsequent mapping process.

```
import pandas as pd

colours_i_like = ['black', 'blue', '#808000', '#483d8b', '#dc143c',
                  '#000080', 'green', '#8b008b', '#ff4500', '#ffa500', '#008856',
                  '#882D17', '#228B22', '#3cb371', '#875692', '#2f4f4f', '#0000ff',
                  '#ff00ff', '#1e90ff', '#db7093', '#ff1493', 'darkgreen',
                  '#ee82ee', '#A1CAF1', 'orange']

colors_df = pd.DataFrame({'basin': basins.MAJ_NAME.unique().tolist(),
                          'colors': colours_i_like})
```

We can then plot the data as before. The command `basins.plot(ax=ax, column='MAJ_NAME', edgecolor='face', color=basins['colors'])` is central to this process. It instructs Python to plot the basins `GeoDataFrame` on predefined axes (ax), utilizing the MAJ_NAME column to identify the major river basins. The *edgecolor='face'* parameter seamlessly blends the borders of each basin with its fill color, eliminating any stark border lines and creating a smooth, cohesive visual. Crucially, the *color=basins['colors']* part of the command applies our custom color scheme, drawn from the colors column in the basins `DataFrame`. This column, a product of merging basin names with our personally curated color palette, assigns a unique hue to each major basin, thus transforming our map into a vivid tapestry of Africa's river systems, both aesthetically pleasing and rich in geographical information.

```
basins = pd.merge(basins, colors_df, left_on='MAJ_NAME', right_on='basin', how='left')

fig, ax = plt.subplots()
basins.plot(ax=ax, column='MAJ_NAME', edgecolor='face', color=basins['colors'])
ax.axis('off')
plt.show()
```

Finally, we can also plot the basins map and add the boundaries of the basins on top to show the breakdown of sub basins within the major basins.

```
fig, ax = plt.subplots()
```

Figure 12.4: Major African river basins.

```
basins.plot(ax=ax, column='MAJ_NAME', edgecolor='face', color=basins['colors'])
basins.boundary.plot(ax=ax, color='black', lw=0.1)

ax.axis('off')
plt.show()
```

Figure 12.5: Major African rivers basins with minor sub basins overlaid.

Combining Rivers and Basins

To achieve our goal of mapping rivers in a way that visually associates them with their corresponding basins, we need to establish a spatial relationship between the river and basin datasets. Using the code: `rivers_basins = gpd.sjoin(africa_rivers, basins, predicate="within")`, we perform a spatial join between the *africa_rivers* and *basins* GeoDataFrames.

The *predicate="within"* argument is key here; it ensures that each river (line geometry) is matched to the basin (polygon geometry) within which it resides. This operation effectively integrates the two datasets, resulting in a new `GeoDataFrame`, *rivers_basins*, where each river is now linked with its respective major basin. This linkage is essential for our subsequent step of coloring the rivers according to the basins they traverse, thereby creating a map that is not only informative in displaying hydrological relationships but also visually coherent in its representation of these natural geographic entities.

```
rivers_basins = gpd.sjoin(africa_rivers, basins, predicate="within")
```

In our final visualization step, we bring the interplay of rivers and their basins to life on the map. This command is instrumental in achieving this:

```
rivers_basins.plot(ax=ax, edgecolor='face',
color=rivers_basins['colors'],
lw=rivers_basins['LW_DIST_UP_KM'])
```

By plotting the *rivers_basins* `GeoDataFrame`, this line ensures that each river is colored according to the major basin it belongs to, as indicated by the *color= rivers_basins['colors']* argument. This color information, previously merged into our `GeoDataFrame`, now serves its purpose of visually associating rivers with their respective basins. The *edgecolor='face'* parameter, as before, ensures the edges of the rivers blend with their color, maintaining a unified and clean visual appeal. Perhaps most interestingly, the `lw=rivers_basins['LW_DIST_UP_KM']` argument dynamically adjusts the line width of each river segment based on its 'LW_DIST_UP_KM' value, which represents a certain characteristic of the river, such as its length or size upstream. This adds an additional layer of depth to our map, allowing viewers to not only distinguish rivers by basin but also understand variations in their physical characteristics. Through this sophisticated yet elegant visualization technique, we create a map that is not only rich in geographical information but also engaging and intuitive to interpret.

```
fig, ax = plt.subplots()
fig.set_size_inches(3.5, 7)

rivers_basins.plot(ax=ax, edgecolor='face', color=rivers_basins['colors'],
                   lw=rivers_basins['LW_DIST_UP_KM'])


ax.axis('off')
plt.show()
```

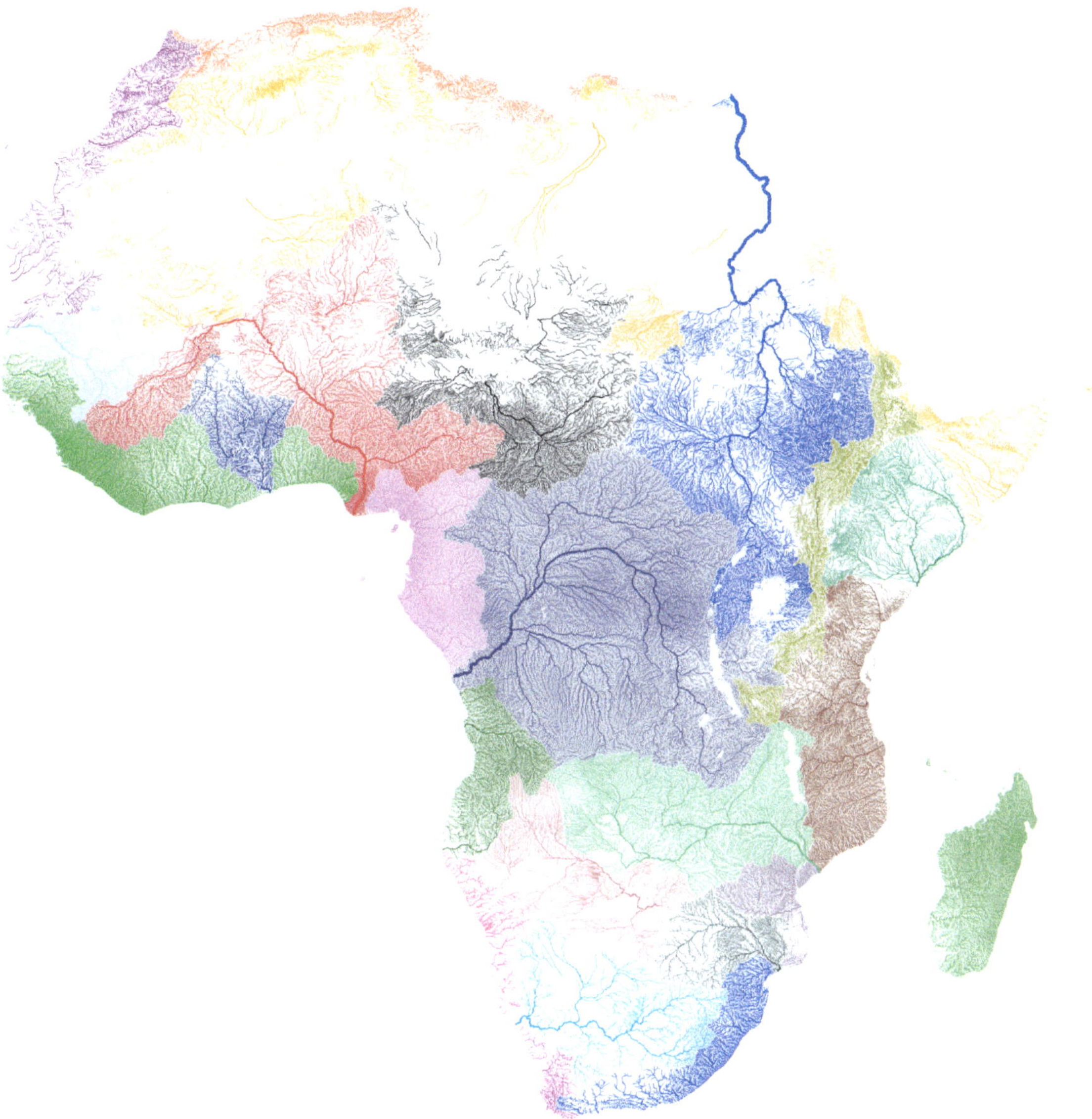

Figure 12.6: African rivers, coloured according to their river basin.

Integrating Topography

In this next phase, we integrate our river map with topographical data to enhance its geographical context. The code snippet provided is for loading and visualizing a topographic raster image, specifically the *ETOPO1_Bed_g_geotiff*. First, we use rasterio, to open the raster file. Then, with `rasterio.plot.show`, we display the first layer of the raster data (denoted by image[0]) as a visual representation.

```
import rasterio
img = rasterio.open('ETOPO1_Bed_g_geotiff.tif')
image = img.read()
```

```
from rasterio.plot import show

show(img)
plt.show()
```

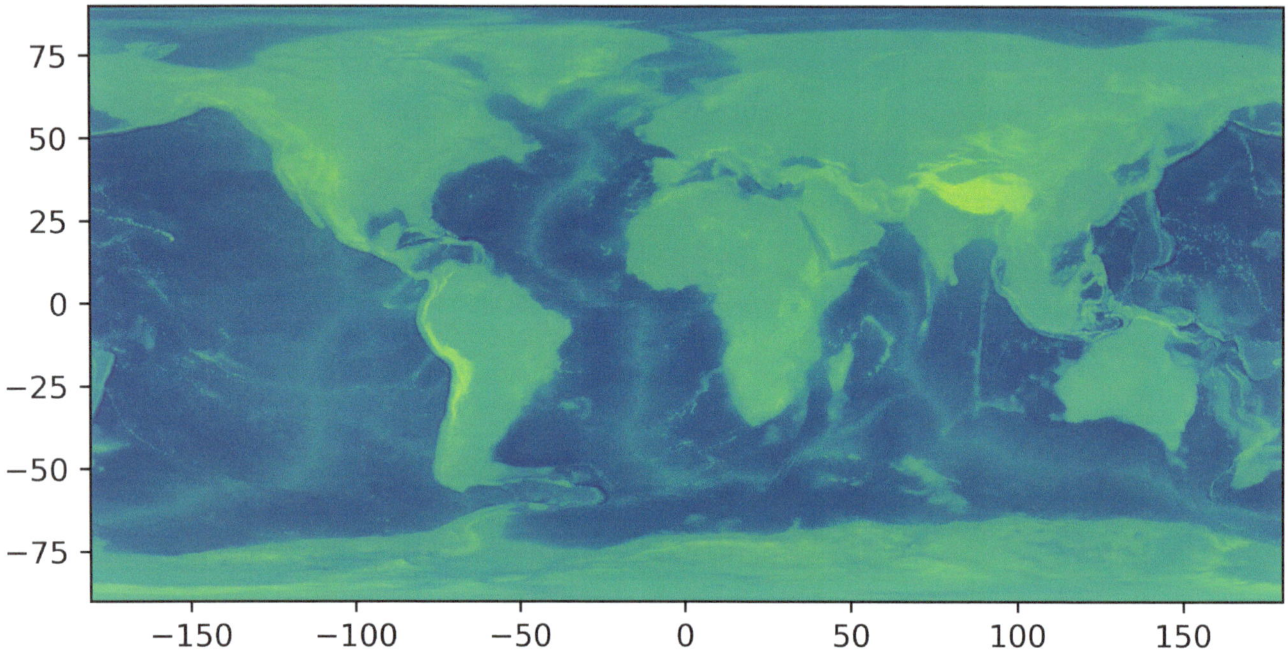

Figure 12.7: Global topography map.

We need to clip our raster to that of African topography. As in previous chapters, we use shapely's mapping function. The mask function from rasterio is used twice. It takes the ETOPO1 raster data (img), and the converted geometries of the African continent, and clips the raster to the extent of these geometries. The *crop=True* argument ensures that the output raster is cropped to the minimal extent of the African geometries, effectively isolating Africa from the rest of the raster data. In the second call to msk.mask, the nodata parameter is set to a value slightly lower than the minimum value in the raster array (*np.amin(a_array[0]) - 1*). This ensure that when we plot the raster, the nodata colours will sit at the bottom of our colourmap.

```
from shapely.geometry import mapping
from rasterio import mask as msk
import numpy as np

a_array, clipped_transform = msk.mask(img,
                              [mapping(geom) for geom in africa.geometry.tolist()],
                              crop=True)
```

```
a_array, clipped_transform = msk.mask(img,
                        [mapping(geom) for geom in africa.geometry.tolist()],
                        crop=True,
                        nodata=(np.amin(a_array[0]) - 1))

value_range = np.amax(a_array) + abs(np.amin(a_array))
```

Our next step is to plot the river basins on top of the topographic map. To effectively visualize the boundaries of the major river basins (identified by MAJ_NAME) on our map, we need to address a challenge. The dataset is made up of small polygons corresponding to the sub-basins. Using the `boundary` method will give us LineStrings corresponding to all of these sub-basins. However, our objective is to display the larger, encompassing boundaries of the major basins. To achieve this, we apply a two-step process in the code.

The code uses a list comprehension to iterate through each unique major basin name (MAJ_NAME) in the basins GeoDataFrame. For each major basin, we select all associated sub-basin geometries (smaller basins within the major basin) and use `unary_union` from the Shapely ops module. This function merges these multiple sub-basin geometries into a single, cohesive geometry representing the entire major basin. The result is a list of unified geometries (*geoms*), each corresponding to one of the major basins.

In the second step, the code iterates over the list of unified major basin geometries. For each unified geometry, it extracts just the boundary, using the *.boundary* attribute. This process strips away the interior details of the polygons, leaving us with clean, concise lines that outline each major river basin.

The outcome of this process is a collection of geometries that accurately represent the boundaries of the major river basins, devoid of the internal subdivisions of the sub-basins. This refined representation is crucial for creating a map that clearly delineates the major river basins, enhancing both the visual appeal and the interpretive clarity of our geospatial visualization.

```
from shapely.ops import unary_union

geoms = [unary_union(basins.loc[basins['MAJ_NAME'] == poly_basin].
         geometry.tolist()) for poly_basin in basins.MAJ_NAME.unique().tolist()]
geoms = [geom.boundary for geom in geoms]
```

We then convert this into a `GeoDataFrame`.

```
new_basins = gpd.GeoDataFrame(
  pd.DataFrame({'Basin': basins.MAJ_NAME.unique().tolist()}), geometry=geoms)
```

We can then seamlessly overlay the basin boundaries onto a topographic map to enhance the visual context of the river systems. The show function from `rasterio.plot` is used to display the clipped topographic raster (*a_array*) with a 'jet' color map, setting the stage on our plot axes (ax).

Subsequently, `new_basins.plot` adds the basin boundaries (previously processed to represent major basins) atop this topographic base, using black lines for clear demarcation.

```
from rasterio.plot import show

fig, ax = plt.subplots()
fig.set_size_inches(3.5, 7)

show(a_array, ax=ax,
     transform=clipped_transform,
     cmap='jet')
new_basins.plot(ax=ax,edgecolor='black', lw=1)

ax.axis('off')
plt.show()
```

The colourmap is not ideal, so the next step is to adjust it to be more appropriate for a topographic baselayer. In crafting a visually compelling topographic map, we adjust the default terrain colormap provided by Matplotlib to better suit our specific needs, particularly focusing on the representation of land features and deliberately omitting water colors. The process begins with extracting a subset of colors from Matplotlib's 'terrain' colormap, which includes a range of colors representing different elevations, including those for water bodies. However, for our purpose, the colors representing water are not required. To achieve this, the code `colors = plt.cm.terrain(np.linspace(0.25, 1, 6279))` is employed. This creates a customized array of colors by sampling the 'terrain' colormap, starting from a point that excludes the lower end of the spectrum (which represents water). The `np.linspace(0.25, 1, 6279)` function specifies that we take 6279 evenly spaced samples from the 25% mark to the end of the colormap, effectively capturing the colors representing land elevations while omitting those for water.

The next step involves setting the initial color in our customized array to white. This is done through `background_colour = np.array([1.0, 1.0, 1.0, 1.0])` and `colors[:1, :] = background_colour`. By changing the first color to white, we ensure that the lowest elevations, or potentially any 'no data' areas, are represented by a clean, neutral background, thus keeping the focus on the topographic features.

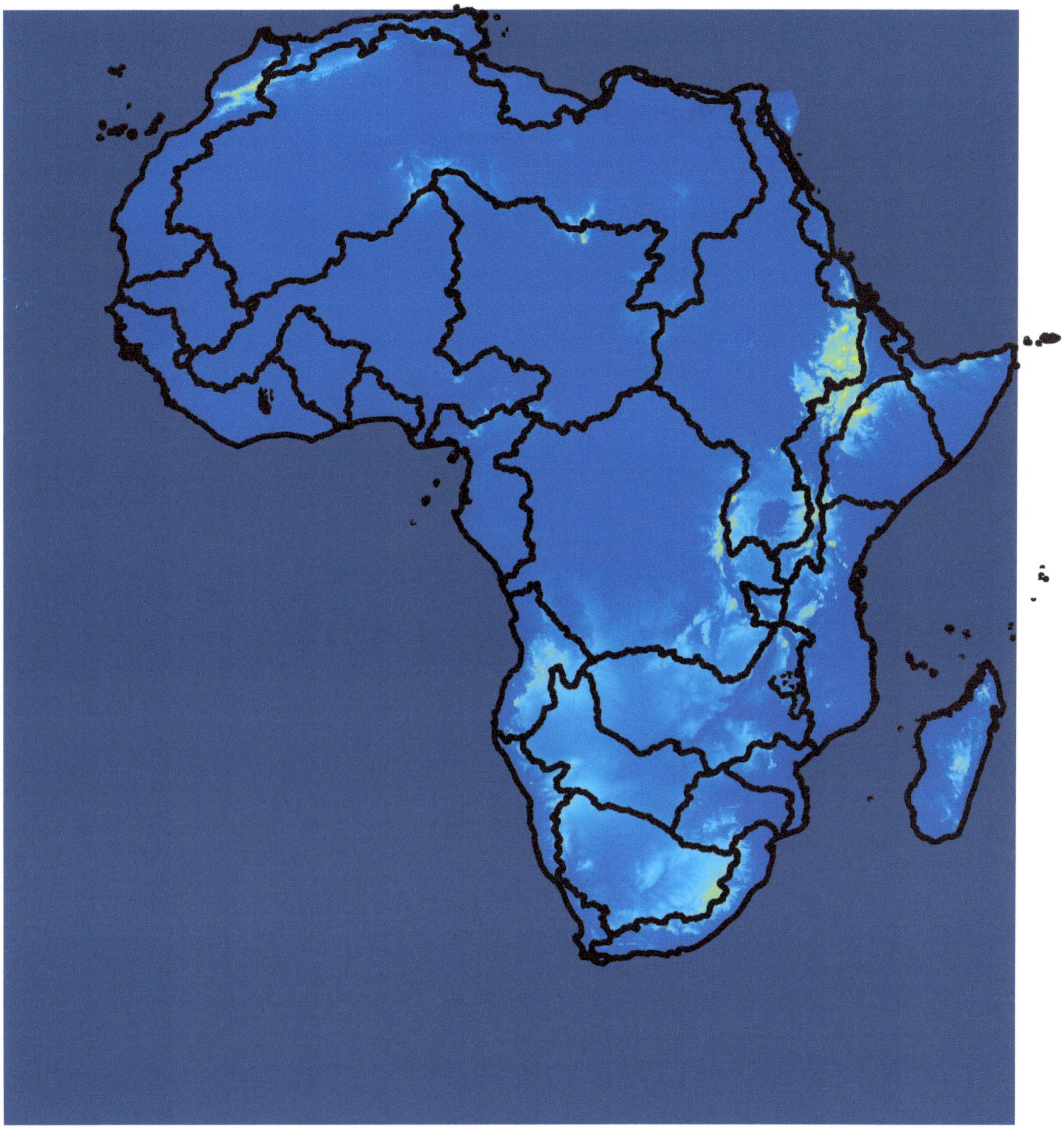

Figure 12.8: African topography with major river basins overlaid.

The adjusted array of colors is then transformed into a usable colormap for Matplotlib visualizations using `our_cmap = ListedColormap(colors)`. This new colormap, *our_cmap*, is tailored specifically for our topographic data, emphasizing land elevations and excluding unnecessary aquatic colorations. Finally, to correctly apply this colormap to our topographic data, we establish a normalization scheme. This is achieved by defining a set of boundaries (`bounds = np.arange(np.amin(a_array), np.amax(a_array)-2500, 10)`) and using `BoundaryNorm(bounds, our_cmap.N)`. The boundaries are determined based on the range of elevation values in our data (*a_array*), and `BoundaryNorm` ensures that these values are accurately mapped to our custom colormap. This normalization is crucial for ensuring that the topographic data is represented in a way that is both visually informative and aligned with the specific elevation ranges of interest.

Through these steps, we create a colormap that is finely tuned for our topographic visualization needs, highlighting land elevations and providing a clear, distinct representation of the geographical terrain.

```
from matplotlib.colors import LinearSegmentedColormap, ListedColormap, BoundaryNorm
from matplotlib import cm

colors = plt.cm.terrain(np.linspace(0.25, 1, 6279))

background_colour = np.array([1.0, 1.0, 1.0, 1.0])
colors[:1, :] = background_colour
our_cmap = ListedColormap(colors)

bounds = np.arange(np.amin(a_array), np.amax(a_array)-2500, 10)
norm = BoundaryNorm(bounds, our_cmap.N)
```

We can then plot the topographic basemap and the basins or the rivers.

```
from rasterio.plot import show

fig, ax = plt.subplots()
fig.set_size_inches(3.5, 7)

show(a_array, ax=ax,
     transform=clipped_transform,
     cmap=our_cmap,
     norm=norm)
new_basins.plot(ax=ax,
                edgecolor='black',
                lw=0.5)
ax.axis('off')
plt.show()
```

```
from rasterio.plot import show

fig, ax = plt.subplots()
fig.set_size_inches(3.5, 7)

show(a_array, ax=ax,
     transform=clipped_transform,
     cmap=our_cmap,
     norm=norm)
rivers_basins.plot(ax=ax, edgecolor='face', color='blue',
                   lw=rivers_basins['LW_DIST_UP_KM'])

ax.axis('off')
```

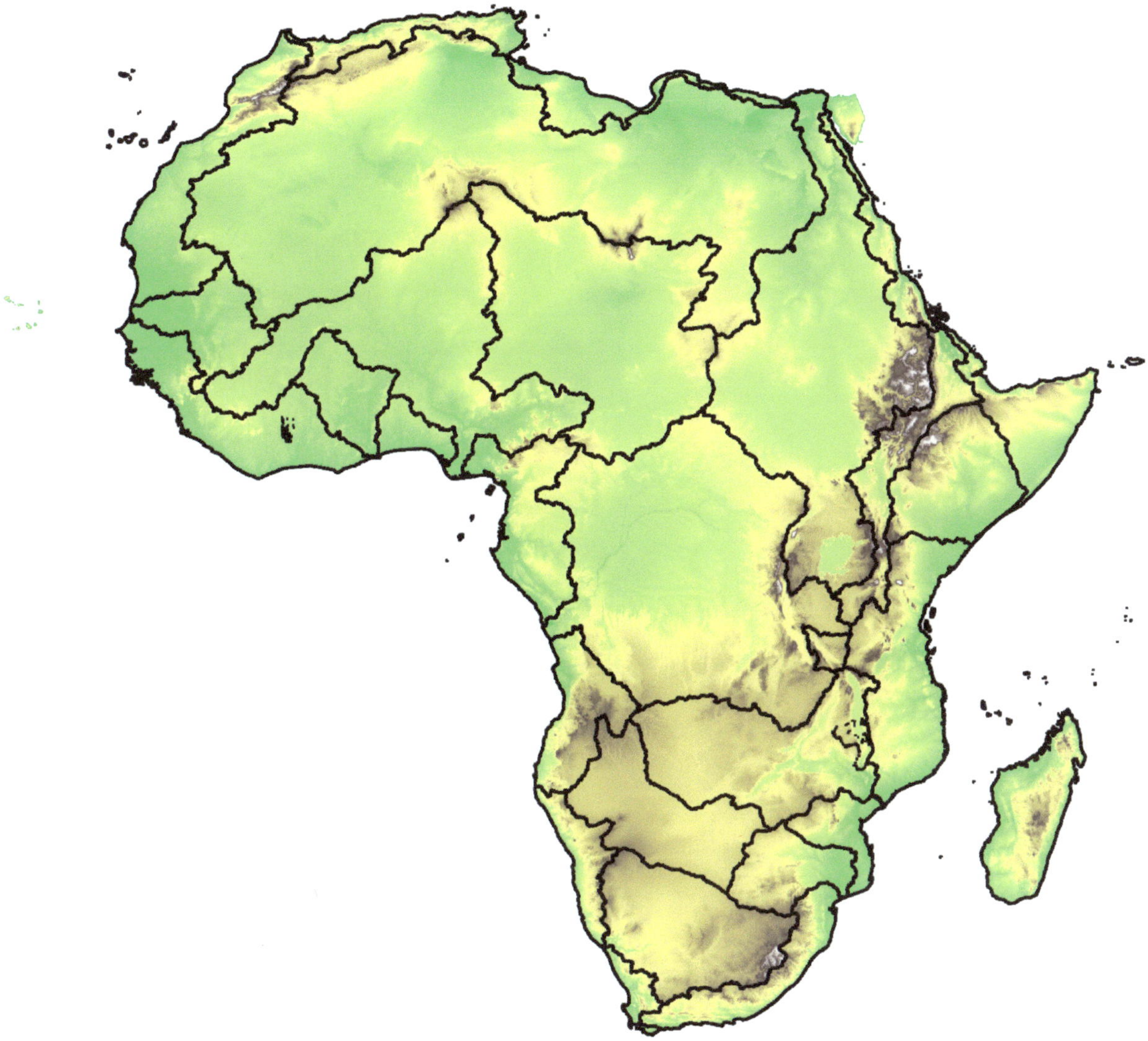

Figure 12.9: African topography with major river basins overlaid and a custom terrain colourmap.

```
plt.show()
```

Finally, in this section, we will explore how to enhance our river maps by visualizing rivers based on their temperature, utilizing the Global River Classification (GloRiC) dataset. GloRiC provides detailed classi-

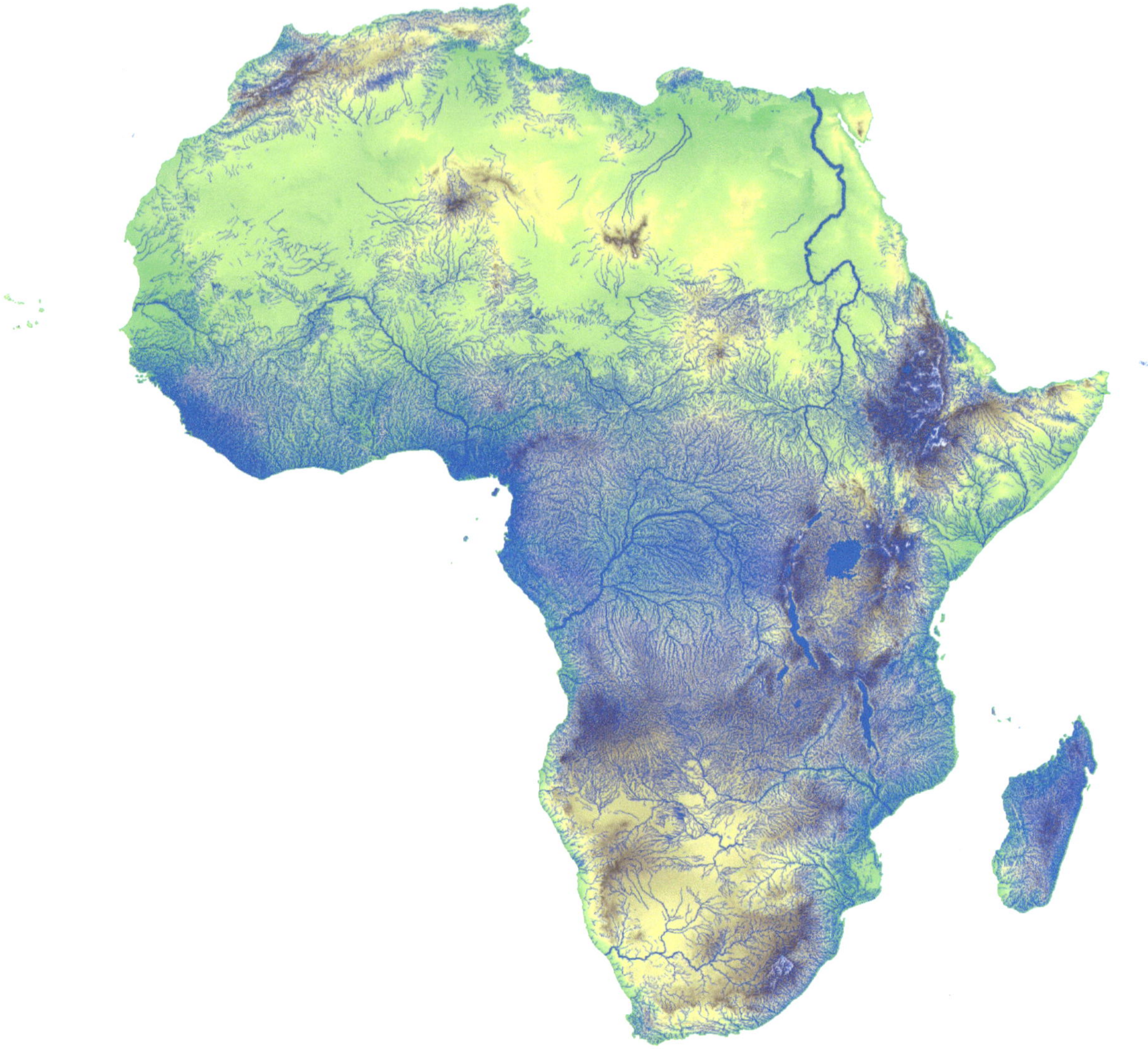

Figure 12.10: African rivers plotted on African topography.

fications for millions of river reaches across the globe, offering insights into hydrologic, climatic, and geomorphic characteristics. By leveraging this dataset, we can enrich our river maps with an additional layer of information—river temperature—adding another dimension to our geospatial analysis.

GloRiC's comprehensive global dataset, which includes 127 distinct river reach types and spans 35.9 million kilometers, allows us to visualize how river temperature varies across different regions and river types. In this section, we will demonstrate a straightforward example: plotting rivers with colors that change according to their temperature. This simple visualization will serve as an introduction to working with the GloRiC dataset.

However, it's important to note that GloRiC offers a wealth of additional data, presenting many other possibilities for exploration. With information on river types, classifications, and environmental characteristics, the potential for deeper analysis is immense. We encourage you to experiment with the dataset and unlock its full potential in your own geospatial projects.

In order to visualize the rivers of Brazil based on their temperature, the first step is to obtain the geographic boundaries of Brazil. We achieve this by filtering the shapefile dataset to extract only the polygon representing Brazil. This will provide us with the spatial context needed to overlay and plot the river data.

As usual we are going to use the natural earth dataset for our shapefiles. The following code filters the country_shapefiles DataFrame, isolating the entry where the 'ADMIN' column corresponds to 'Brazil.' With this, we now have the geographic boundaries of Brazil, which will serve as the backdrop for plotting its river systems, classified by temperature.

This setup enables us to map only the rivers within Brazil, ensuring that the visualization is region-specific and relevant to our goal of analyzing river temperatures.

```
brazil = country_shapefiles.loc[country_shapefiles['ADMIN'] == 'Brazil']
```

To visualize the river temperatures in Brazil, we next load the GloRiC dataset, which contains extensive river classification information for South America. This is a very large dataset, containing millions of river reaches, and can take a significant amount of time to load and process, especially on older laptops.

This step will load the GloRiC dataset as a GeoDataFrame, enabling us to work with the temperature data of rivers within the region.

```
gloric = gpd.read_parquet("GloRiC_v10.shp")
```

Once we have the GloRiC dataset and Brazil's shapefile, we perform a spatial join to isolate only the rivers within Brazil. This allows us to focus our visualization on the Brazilian river network.

Since the spatial join operation has been covered in previous chapters, we won't dive into the details here. This step simply ensures that we are selecting the rivers located within the boundaries of Brazil, enabling us to map the temperature data specifically for this region.

```
brazil_rivers = gpd.sjoin(gloric, brazil, op="within")
```

In this segment, we're scaling the river widths based on the Log_Q_avg attribute, similar to how we scaled river widths earlier in the chapter. The process is the same, but it's important to note the specific scaling factors used here: 0.6 and 0.01. These values define the range for the line widths, with 0.6 as the maximum and 0.01 as the minimum.

As discussed earlier in the chapter, it's often necessary to tweak these scaling values depending on the specific map you're making and the dataset you're working with. In this case, adjusting the values ensures that the river widths are visually appropriate for representing the variation in flow, but you may need to experiment with different ranges for other contexts.

```
import pandas as pd
import numpy as np

leftSpan = np.amax(brazil_rivers['Log_Q_avg']) - np.amin(brazil_rivers['Log_Q_avg'])
rightSpan = 0.6 - 0.01
valueScaled = (brazil_rivers['Log_Q_avg'] - np.amin(brazil_rivers['Log_Q_avg'])) / leftSpan
brazil_rivers['LW_Log_Q_avg'] = 0.01 + (valueScaled * rightSpan)
```

Next, we extract the upper and lower temperature limits for each river by examining the Temp_min attribute in the dataset. This attribute represents the minimum temperature of each river reach, which we will use to color the river lines.

The following code prints the minimum and maximum values of Temp_min, helping us understand the temperature range across all the rivers in Brazil.

By identifying these temperature limits, we can appropriately map a color gradient to represent the rivers, where cooler temperatures are depicted with one color and warmer temperatures with another. This ensures that the visual representation accurately reflects the temperature variations across Brazil's rivers.

```
print(np.amin(brazil_rivers['Temp_min']))
print(np.amax(brazil_rivers['Temp_min']))
```

```
9.9
27.2
```

As we covered in a previous chapter, we will now set up a color map to visualize river temperatures. Here, we use the jet colormap and adjust it to match the temperature range of the dataset.

This snippet creates a custom colormap (our_cmap) using 190 color levels and sets the temperature range from 10°C to 27°C with intervals of 0.1°C. The BoundaryNorm ensures the colors are properly distributed across the temperature values, allowing us to color the rivers dynamically based on their minimum temperature (Temp_min).

```
from matplotlib import cm
from matplotlib.colors import ListedColormap
from matplotlib import colors

our_cmap = cm.get_cmap('jet', 190)
our_cmap = our_cmap(np.linspace(0, 1, 190))
our_cmap = ListedColormap(our_cmap[1:])
bounds = np.arange(10, 27, 0.1)
norm = colors.BoundaryNorm(bounds, our_cmap.N)
```

In this final piece of code, we visualize the Brazilian rivers on a map, with the rivers colored according to their temperature and scaled by their flow rates. The first line, `fig, ax = plt.subplots(facecolor='black')`, creates a figure and axes object with a black background, setting the stage for the plot. This gives us a clean, contrasting background for the colored rivers to stand out.

Next, we plot the rivers with brazil_rivers.plot(), where the width of each river is determined by its flow (`lw=brazil_rivers['LW_Log_Q_avg']`), and the color represents the minimum temperature (`column=brazil_rivers['Temp_min']`). The *cmap=our_cmap* applies the custom colormap we previously defined, and norm=norm ensures the colors are properly normalized according to the temperature range.

To add a color bar, we first create a new axes object with `fig.add_axes([0.3, 0.21, 0.14, 0.01])`, which positions and sizes the color bar within the figure. The ScalarMappable connects the colormap and normalization, preparing it for use in the color bar. Then, `fig.colorbar()` creates the actual color bar, setting it to be horizontal, with specific tick marks representing temperatures from 12°C to 28°C. We ensure that the color bar does not have any edges (drawedges=False) and format the tick labels with `format='%.3g'` for a clean, professional appearance.

Finally, we customize the appearance of the color bar. The tick labels are explicitly set to match the temperature ticks, and the outline of the color bar is removed with `cb.outline.set_visible(False)` to maintain a minimalistic design. We further adjust the tick label sizes and appearance with `cb.ax.tick_params()`, ensuring that the labels are clear, small, and white to fit the black background. Lastly, the label "Temperature Celsius" is added to the color bar for clarity. The axes themselves are turned off with `ax.axis('off')`, ensuring that only the rivers and temperature data are highlighted in the final visualization.

```

fig, ax = plt.subplots(facecolor='black')

brazil_rivers.plot(ax=ax, lw=brazil_rivers['LW_Log_Q_avg'],
                   column=brazil_rivers['Temp_min'], cmap=our_cmap, norm=norm)

fig = ax.get_figure()
cax = fig.add_axes([0.3, 0.21, 0.14, 0.01])
sm = plt.cm.ScalarMappable(cmap=our_cmap, norm=norm)
sm._A = []
cb = fig.colorbar(sm, cax=cax, orientation="horizontal", pad=0, format='%.3g',
                  ticks=[12, 16, 20, 24, 28], drawedges=False)
cb.ax.set_xticklabels(['12', '16', '20', '24', '28'])
cb.outline.set_visible(False)
cb.ax.tick_params(labelsize=4, width=0.2, length=0.75, color='white')
cbytick_obj = plt.getp(cb.ax, 'xticklabels' ) #Set y tick label color
plt.setp(cbytick_obj, color='white')
cb.ax.set_xlabel('Temperature Celcius', fontsize=5, color='white', labelpad=-20)
ax.axis('off')

plt.show()
```

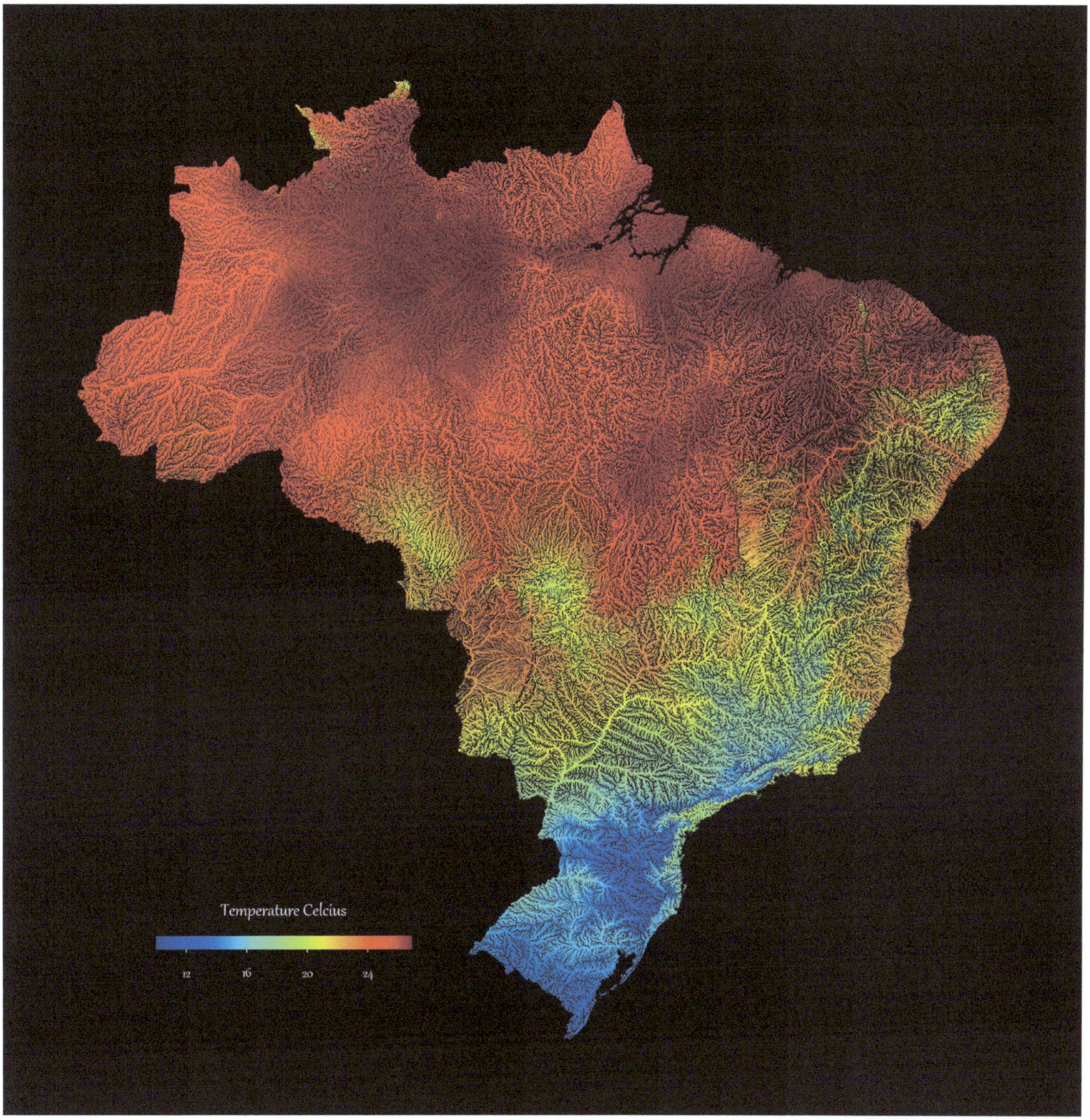

Figure 12.11: Minimum river temperature in Brazil.

13

Mapping Lighthouses

13.1 Lighthouse Maps

In this chapter, we embark on a journey that marries the precision of geospatial data manipulation with the artistry of visual storytelling. Our focus shifts to the rugged and picturesque coastlines of the British Isles, where lighthouses have stood as beacons of safety and symbols of human ingenuity for centuries. Leveraging the extensive and dynamic dataset of OpenStreetMap, we will meticulously extract the locations of lighthouses across this region, setting the stage for a unique cartographic endeavor.

Our objective transcends the mere plotting of points on a map. Instead, we delve into advanced data manipulation techniques to create a visualization that not only marks the lighthouses but also artistically simulates the way their lights shine out to sea. This approach transforms a standard geographical representation into a vivid portrayal, capturing the essence of these lighthouses as they cut through the darkness and guide mariners through the treacherous waters.

This chapter is more than a technical walkthrough; it's an invitation to think creatively about geospatial data and its potential to tell stories. Through our exploration, we will not only enhance our technical skills in data handling and visualization but also develop an appreciation for the narrative power of maps. As we plot these symbols of hope and guidance, we will illuminate not just the British Isles' coastlines, but also the broader horizons of what can be achieved with advanced data visualization techniques.

In this initial step, we load the geographical boundary data for the

British Isles. Using the GeoPandas library, we read a shapefile that contains country boundaries from NaturalEarth. The shapefile is accessed using `gpd.read_file()`, loading the data into a `GeoDataFrame` called *gdf*. We then filter this data to include only the areas of interest — Ireland, the United Kingdom, and the Isle of Man — by selecting rows where the 'ADMIN' column matches these names. The result is stored in the GeoDataFrame bi, which now represents the geographic boundary of the British Isles, forming the foundational geographic context for our subsequent analyses and visualizations.

```
import geopandas as gpd

gdf = gpd.read_file('NaturalEarth/10m_cultural/ne_10m_admin_0_countries.shp')
bi = gdf.loc[gdf['ADMIN'].isin(['Ireland', 'United Kingdom', 'Isle of Man'])]
```

To retrieve data for lighthouses we are going to use Overpy. Overpy is a Python wrapper for the Overpass API, which is used to query OpenStreetMap (OSM) data. Overpy makes it easier to interact with the Overpass API using Python, allowing you to run queries and retrieve data from OSM in a more Pythonic way.

Overpass API itself is a read-only API that allows users to extract custom data from the vast and continuously updated OpenStreetMap database, based on specific criteria, such as geographic location, type of object, tags, and more. This can include a wide range of data, from roads and buildings to natural features and points of interest like lighthouses.

Using Overpy, you can write queries in the Overpass Query Language and send these queries to the API. The library then parses the API's response and provides you with Python objects that are easier to work with. This functionality is particularly useful for projects involving detailed geographic analyses and custom map visualizations, such as the one you're undertaking in your chapter.

This code snippet demonstrates how to use Overpy, a Python library for querying OpenStreetMap (OSM) data, to retrieve the locations of lighthouses in the British Isles.

The code begins by importing necessary libraries: overpy for querying OSM data, os for interacting with the operating system, and pandas for data manipulation. The Overpass API endpoint is defined as overpass_url. `bounds = bi.total_bounds` extracts the geographic boundaries (latitude and longitude) of the British Isles from the previously loaded `GeoDataFrame` called *bi*. The *total_bounds* attribute returns a bounding box that encompasses all geometries within the `GeoDataFrame`.

`api = overpy.Overpass(retry_timeout=True)` creates an instance of the Overpass API with retry functionality enabled. The `api.query(...)` function sends a query to the Overpass API. The query is formatted to retrieve nodes tagged as *man_made=lighthouse* within the defined bounds. The bounds are specified in the order: minimum latitude, minimum longitude, maximum latitude, maximum longitude. lat and lon lists are populated with the latitude and longitude of each node (lighthouse) returned in the query result. `df = pd.DataFrame({"lat": lat, "lon": lon})` creates a `DataFrame` using pandas, where the latitude and longitude values are stored in columns named 'lat' and 'lon', respectively.

By executing this code, we effectively gather and organize the geographical locations of lighthouses across the British Isles into a structured pandas `DataFrame`, laying the groundwork for subsequent spatial analysis and visualization of these iconic coastal features.

```
import overpy
import os
import pandas as pd
overpass_url = "http://overpass-api.de/api/interpreter"

bounds = bi.total_bounds
api = overpy.Overpass(retry_timeout=True)
result = api.query(f"""node["man_made"="lighthouse"](
                    {bounds[1]},{bounds[0]},{bounds[3]},{bounds[2]});out;""")
lat = [float(node.lat) for node in result.nodes]
lon = [float(node.lon) for node in result.nodes]
df = pd.DataFrame({"lat": lat, "lon": lon})
```

We can then convert our latitude and longitude values into a `GeoDataFrame` and plot the data as we have done in previous chapters.

```
from shapely.geometry import Point

geometries = [Point(long, lat) for long, lat in zip(lon, lat)]
lh = gpd.GeoDataFrame(df, crs='EPSG:4326', geometry=geometries)
```

```
import matplotlib.pyplot as plt

fig, ax = plt.subplots()
lh.plot(ax=ax)
plt.show()
```

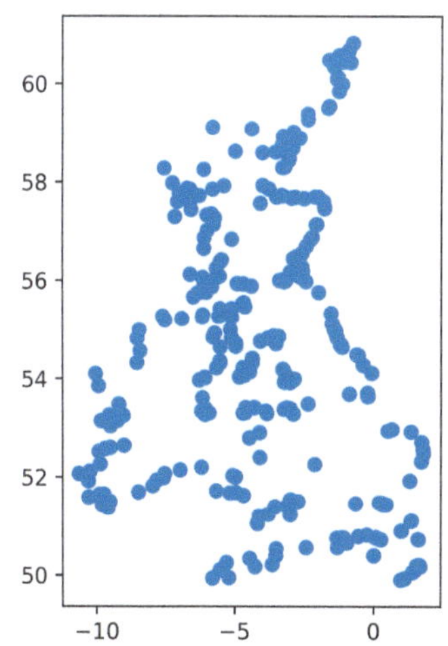

Figure 13.1: Locations of lighthouses in the British Isles.

Now we want to plot these points and simulate light emitted. Our strategy involves a creative manipulation of our lighthouse dataset, where we'll duplicate the data multiple times, each time adding an incrementally larger buffer around the points. This buffer symbolizes the light emitted, growing fainter with distance. By layering these buffered datasets and applying a consistent transparency (alpha value) to each layer, we create a visually striking effect. The intensity of the light will appear stronger near the lighthouse due to overlapping buffers, and gradually fade as the distance increases, where fewer overlaps occur.

This code snippet is designed to implement the shining effect for the lighthouses on the map, as described earlier.

This list comprehension creates a series of dictionaries:

```
[{'geom': lh.geometry.buffer(x).geometry.tolist(),
  'buffer': np.zeros(lh.shape[0]) + x}
  for x in np.arange(0.1, 0.3, 0.01)]
```

Each dictionary corresponds to a different buffer size, ranging from 0.1 to 0.3 (with an increment of 0.01) around each lighthouse point (`lh.geometry.buffer(x)`). For each buffer size x, the code generates a buffered geometry around each lighthouse and creates a corresponding numpy array filled with the buffer size *x* (using: `np.zeros(lh.shape[0]) + x`). This way, each buffered geometry is associated with its buffer size.

`new_geoms = pd.concat([pd.DataFrame(df) for df in new_geoms])` concatenates all the generated DataFrames (each representing different buffer sizes) into a single `DataFrame`, *new_geoms*. This `DataFrame` now contains all the buffered geometries and their respective buffer sizes.

`new_geoms = gpd.GeoDataFrame(new_geoms, crs='EPSG:4326', geometry=new_geoms['geom'])` converts the combined `DataFrame` into a `GeoDataFrame` to facilitate spatial operations. The coordinate reference system (CRS) is set to 'EPSG:4326', which is commonly used for geographic coordinates.

`new_geoms['alpha'] = 0.024` adds a new column, alpha, to the `GeoDataFrame` and assigns a constant value of 0.024 to it. The alpha value controls the transparency of the geometries when plotted. A consistent alpha value ensures that the intensity of the shining effect diminishes with distance from the lighthouse, as the overlap of buffered geometries decreases.

```
import numpy as np

new_geoms = [{'geom': lh.geometry.buffer(x).geometry.tolist(),
              'buffer': np.zeros(lh.shape[0]) + x} for x in np.arange(0.1, 0.3, 0.01)]

new_geoms = pd.concat([pd.DataFrame(df) for df in new_geoms])
new_geoms = gpd.GeoDataFrame(new_geoms, crs='EPSG:4326', geometry=new_geoms['geom'])
```

When we come to plot the data we are doing a few things:

- We are setting the background colour to black because shiny things look better in the dark.

- We are plotting our original lighthouse dataset.
- We are plotting our buffered geometries. We are setting a single alpha value.

```
fig, ax = plt.subplots(facecolor='black')
ax.patch.set_facecolor('black')
lh.plot(ax=ax, markersize=0.1, alpha=1, edgecolors='none', color='white')
new_geoms.plot(ax=ax, alpha=0.024, edgecolors='none', color='white')
plt.show()
```

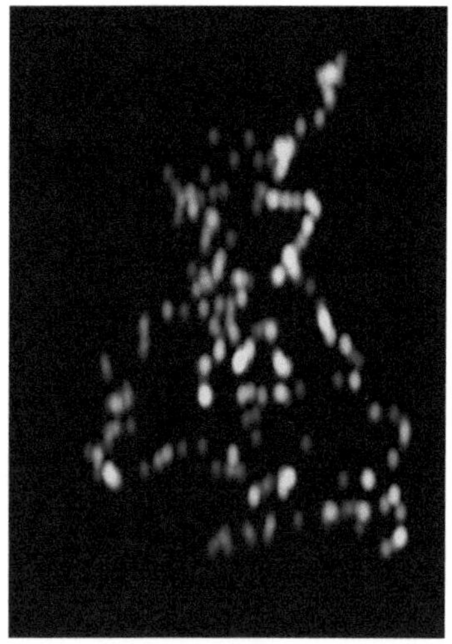

Figure 13.2: Lighthouses in the British Isles plotted with decreasing brightness away from shore.

Lighthouses do shine onto land but we only care about how they shine out to sea. So we are going to cut out the parts of our polygons which intersect with the land. To achieve this, we employ a sophisticated geometric operation using the overlay function in GeoPandas. We apply the overlay function with the *how='difference'* parameter to the buffered geometries in *new_geoms* (representing the expanded reach of each lighthouse's light) and the land boundaries in *bi* (representing the British Isles). This operation effectively "carves out" the sections of the light buffers that intersect with the landmass, leaving us with an accurate representation of the lighthouse beams as they would appear shining out to sea. The result of this operation, stored in the difference `GeoDataFrame`, captures the essence of lighthouses fulfilling their purpose - guiding seafarers with their beams that cut across the waters, unimpeded by the land. This nuanced approach adds a layer of realism to our map, visually conveying how these beacons of light interact with the coastal landscapes of the British Isles.

```
difference = new_geoms.overlay(bi, how='difference')
```

Plotting the data shows off our lighthouse map.

```
fig, ax = plt.subplots(facecolor='black')
ax.patch.set_facecolor('black')
lh.plot(ax=ax, markersize=0.1, alpha=1, edgecolors='none', color='white')
difference.plot(ax=ax, alpha=0.024, edgecolors='none', color='white')
plt.show()
```

Figure 13.3: Lighthouses in the British Isles plotted with decreasing brightness away from shore.

14

Mapping the World's Shipping Lanes

14.1 Points to Raster

In the dynamic world of geospatial data analysis, the ability to fluidly transform data from one form to another unlocks a multitude of possibilities for understanding and interpreting spatial relationships. This chapter delves into the fascinating process of converting geospatial data between two common formats: points (latitude-longitude coordinates) and raster.

Our journey begins by exploring how a collection of latitudinal and longitudinal coordinates, representing tangible locations, can be translated into a raster format. To bring this concept to life, we will use a dataset of restaurants and bars across Europe as our guide. This practical example will illustrate the process of converting point data into a raster grid but also highlight the nuances and decisions involved in such a transformation, including the selection of appropriate grid sizes and resolution.

Next, we pivot to a seemingly reverse yet equally intriguing process: transforming raster data back point format. This part of the chapter will utilize shipping lanes in Northern Europe as a case study. Through this example, we will explore how raster data, often used for its ability to represent continuous spatial phenomena, can be effectively converted into discrete points. This conversion is crucial for specific applications where point data is more meaningful or required for further analysis.

Throughout the chapter, we will employ Python, a powerful and versatile programming language, as our tool of choice. Python's rich ecosystem of geospatial libraries and frameworks makes it an ideal platform for performing these transformations. We will leverage libraries such as GDAL, rasterio, and geopandas, demonstrating their functionality and ease of use in handling complex geospatial data operations.

By the end of this chapter, you will have a comprehensive understanding of how to convert between point and raster data formats using Python. This knowledge will not only enhance your skillset in geospatial data handling but also broaden your perspective on the potential applications and implications of these transformations in the real world.

Points to Raster

In this chapter, we'll be delving into the conversion of geospatial data between point and raster formats. It's important to note that the process of querying OpenStreetMap to gather latitude and longitude data for restaurants and bars has been covered in a previous chapter. For our current exercises, we will bypass this step and directly utilize a pre-prepared CSV file. This file, which I have compiled beforehand, contains the relevant geospatial data needed for our analysis. This approach allows us to focus more closely on the techniques of transforming this data into raster format and vice versa, using Python's powerful geospatial libraries.

```
import glob
import pandas as pd

dataset = pd.read_csv("bars.csv")

print(dataset)
```

```
          lat        lon
0         39.789151  19.920981
1         41.048885  20.634035
2         41.111911  20.023851
3         41.135273  20.385996
4         39.651711  19.829637
...          ...        ...
87340      41.552429  -8.423668
87341      42.614656  -8.775682
87342      42.615154  -8.775323
87343      46.662414   0.360084
87344      50.722899  -3.531709

[1334940 rows x 3 columns]
```

In this section of our analysis, we utilize the numpy library, a fundamental package for scientific computing in Python, to create a two-dimensional histogram from our geospatial dataset. The code import numpy as np brings the numpy library into our working environment, allowing us to leverage its powerful array and numerical operations.

The core of this operation is:
`np.histogram2d(dataset['lat'], dataset['lon'], bins=400)`. This function generates a two-dimensional histogram, a type of plot that shows the frequency distribution of two variables simultaneously. In our case, these variables are the latitude (lat) and longitude (lon) coordinates of the locations in our dataset. By specifying bins=400, we divide each axis (latitude and longitude) into 400 intervals, creating a grid of 400x400 bins over our data range.

The result, stored in hist, is a two-dimensional array where each element represents the count of points (in this instance, restaurants and bars) that fall into each bin of the grid. This histogram effectively transforms our point data into a raster-like format, where the frequency of data points within each geographic area is represented as a value in a matrix. This matrix can then be used for further analysis or visualization, providing insights into the spatial distribution and density of the points across the region.

```
import numpy as np

hist = np.histogram2d(dataset['lat'], dataset['lon'], bins=400)
```

Flipping the Array Upside Down: The function `np.flipud()` is then applied to this 2D array. `np.flipud` stands for "flip up down" and is a numpy function used to flip arrays vertically (upside down). The reason for this operation is rooted in the way image data is typically handled in computational environments. In many image processing contexts, the origin (0,0) is considered to be the top-left corner of the image. However, in mathematical plotting (and consequently in histograms like ours), the origin is often at the bottom-left corner. Flipping the array upside down aligns the data with the image processing standard, ensuring that the spatial orientation of the data (especially when visualized) matches our real-world expectations.

```
array = np.flipud(hist[0])
```

In a previous chapter, we explored custom colormap creation using Matplotlib. This revised code snippet continues with the theme of customizing colormaps for data visualization. The process begins by importing `ListedColormap` from *matplotlib.colors* and *cm* from *matplotlib*, which are key to working with colormaps. The code defines a background color as fully transparent black, represented by *background_colour*. A reversed blue colormap (*Blues_r*) with 101 shades is then created using `cm.get_cmap()`, and stored in the variable *blues_cmap*. The colors from this colormap are retrieved and stored in *blues_colors*. The first color in this array is replaced by the defined *background_colour* to introduce transparency. Finally, this modified array of colors is used to create a new `ListedColormap` object, named *newcmp*, which incorporates the gradation of blue shades against a transparent background. This colormap can be applied in various visualizations to represent data with nuanced blue tones, enhanced by the transparency at its base.

```
from matplotlib.colors import ListedColormap
from matplotlib import cm

background_colour = np.array([0.0, 0.0, 0.0, 0.0])
blues_cmap = cm.get_cmap('Blues_r', 101)
blues_colors = blues_cmap(np.linspace(0, 1, 101))
blues_colors[:1, :] = background_colour
newcmp = ListedColormap(blues_colors)
```

As in previous chapters, we could use Matplotlib's `imshow` function or Rasterio's `show` function to visualize geospatial data but as this has been covered I will leave that up to you to try. This code snippet takes a different approach, using Matplotlib and Cartopy for advanced geospatial plotting.

A figure object is created with a black background (*facecolor='black'*). An axes object is created with the *projection=ccrs.Robinson()* argument. This specifies that the Robinson projection is to be applied to the plot. The `transform_points` method is used to transform the longitude and latitude points from the dataset into the Robinson projection coordinate system. This transformation is crucial for accurate plotting on the map projection. The transformed points are then plotted using `ax.hist2d`, which creates a 2D histogram on the axes. The histogram bins are set to 800, and a logarithmic normalization (`colors.LogNorm()`) is applied for color scaling, enhancing the visibility of data across a wide range of values. The colormap newcmp", created earlier, is used for the color scheme of the histogram.

```
import matplotlib.pyplot as plt
from matplotlib import colors
from cartopy import crs as ccrs

fig = plt.figure(facecolor='black')
fig.set_size_inches(7, 3.5)
ax = plt.axes(projection=ccrs.Robinson())
xynps = ax.projection.transform_points(ccrs.Geodetic(), dataset['lon'].to_numpy(),
                                        dataset['lat'].to_numpy())

ax.hist2d(xynps[:,0], xynps[:,1], bins=600, norm=colors.LogNorm(), cmap=newcmp)
ax.patch.set_facecolor('black')
ax.axis('off')
plt.show()
```

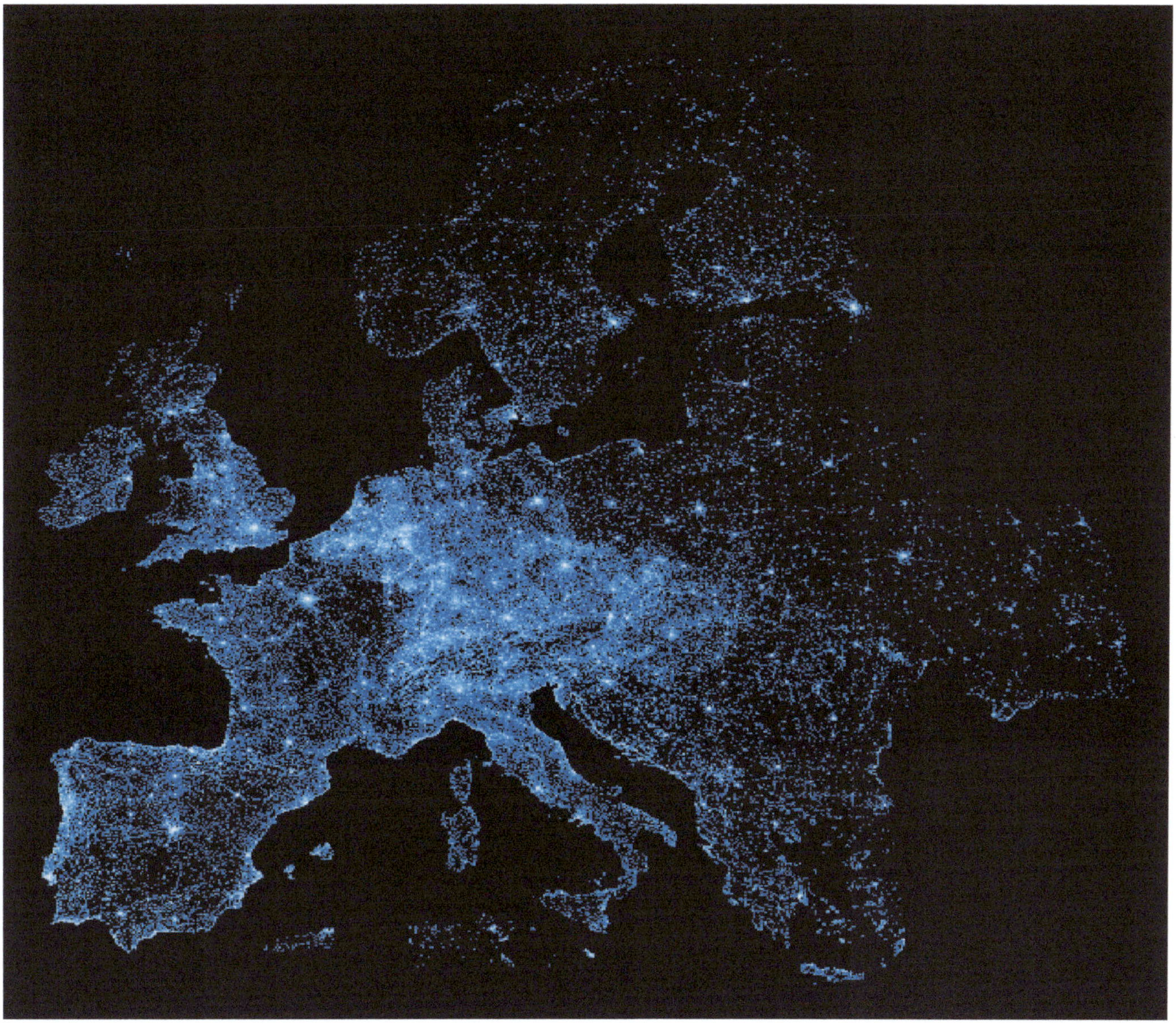

Figure 14.1: Bars and restaurants of Europe

Raster to Points

In this next step, we begin working on the shipping lanes example by loading the necessary data using Rasterio. We start by loading a raster file named "shipdensity_global.tif". This file contains global shipping lane density data in TIFF format.

```
import rasterio

src = rasterio.open("shipdensity_global.tif")
```

In this stage of the analysis, we are preparing to clip the shipping lanes array to focus specifically on the North Sea area. This is achieved by creating a Geopandas `GeoDataFrame`, which will represent a geometric boundary around the North Sea. First, A `GeoDataFrame` named *north_europe* is created using Geopandas, initialized with a single data row named 'north_europe'. The coordinate reference system (CRS) is set to 'EPSG:4326'.

The geometry for this `GeoDataFrame` is initially defined as a `Point`, positioned at latitude 55.8566 and longitude 0.3522. This location is chosen as it approximately marks a central point in the North Sea. To create an area representing the North Sea, this point is buffered using the `buffer` method, which expands the point into a square region with a 15-degree radius. The buffer is formed using one segment to approximate the circle, providing a basic but effective representation.

Finally, to better align with the North Sea's geographical orientation, this square geometry is rotated by 45 degrees using the rotate method. The result is a geometric shape in the north_europe `GeoDataFrame` that approximates the area of the North Sea, ready to be used for clipping the raster data of the shipping lanes, allowing for a more focused and region-specific analysis.

```
import geopandas as gpd
from shapely.geometry import Point

north_europe = gpd.GeoDataFrame(pd.DataFrame({'name': ['north_europe']}),
                                crs='EPSG:4326',
                                geometry=[Point(0.3522, 55.8566)])
north_europe.geometry = north_europe.geometry.buffer(15, 1)
north_europe.geometry = north_europe.geometry.rotate(45)
```

We can then use the `rasterio.mask.'mask function to clip the raster data ('src)` to the area defined by the *north_europe* `GeoDataFrame`. The mapping function from Shapely converts the geometries in *north_europe* to a format compatible with Rasterio's masking operation. The *crop=True* parameter ensures that the output array (*clipped_array*) is cropped to the extent of the clipping mask, and *nodata=0* sets the pixel value for areas outside the mask to zero. The result is a clipped version of the original raster data, now focused on the North Sea region, along with its corresponding affine transform (*clipped_transform*). This operation is crucial for isolating the area of interest for more detailed analysis.

```
from rasterio import mask as msk
from shapely.geometry import mapping

clipped_array, clipped_transform = rasterio.mask.mask(src, [
    mapping(geom) for geom in (north_europe.geometry.tolist())
], crop=True, nodata=0)
```

To visualize the clipped raster data of the shipping lanes in the North

Sea region, we can use the imshow function. In this context, it will be used to plot the *clipped_array* — the array containing the shipping lane data that has been clipped to the North Sea.

```
# Plot the raster data
plt.imshow(clipped_array[0], norm=colors.LogNorm(), cmap='viridis')
plt.axis('off')
plt.show()
```

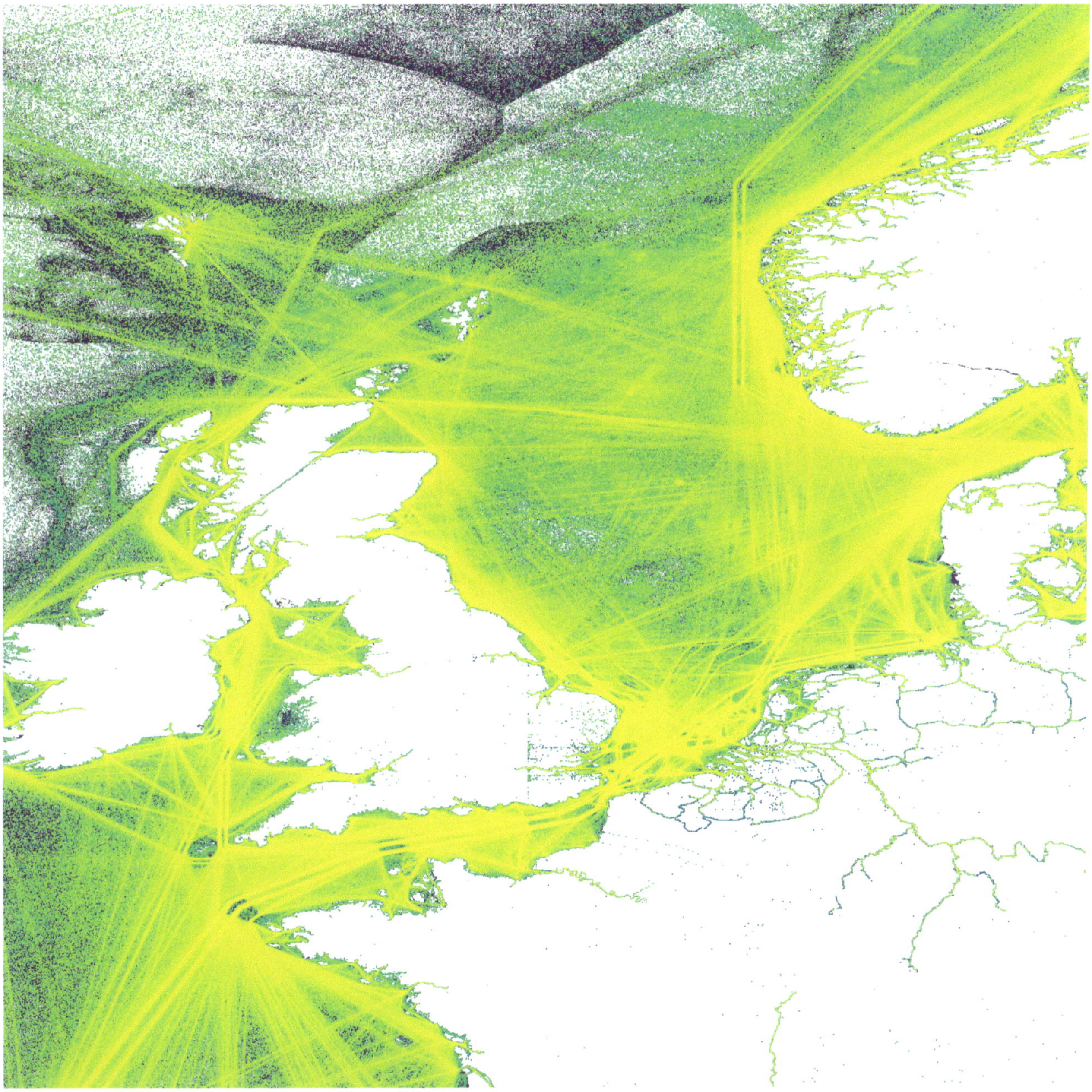

Figure 14.2: Shipping lanes of Northern Europe

In this stage of the analysis, we are transitioning from raster to vector data by converting our clipped raster of shipping lanes into a geospatial dataframe, where each raster pixel is represented as a point with corresponding geographic coordinates.

The process begins by determining the dimensions of the clipped raster array, specifically the number of rows and columns, which represent the pixel count vertically and horizontally. This is achieved through the shape attribute of the clipped array. With this information, a grid of pixel indices is generated using NumPy's meshgrid function. This function creates a coordinate matrix for the columns and rows, effectively mapping out the position of each pixel in the raster.

These pixel indices are then transformed into real-world geographic coordinates. The `rasterio.transform.xy` function is employed to convert the row and column indices into x (longitude) and y (latitude) coordinates. This transformation relies on the *clipped_transform* obtained from the raster clipping process, ensuring that each pixel is accurately mapped to its geographic location.

Following the coordinate transformation, a Pandas `DataFrame` is constructed. This `DataFrame` contains the x and y coordinates along with the corresponding values extracted from the raster, where each value represents a data point from the shipping lane density. Thus, each row in the DataFrame correlates to a specific point in the raster.

The final step is to convert this `DataFrame` into a `GeoDataFrame`. This is accomplished using Geopandas, where the `gpd.points_from_xy` function creates `Point` geometries from the x and y columns. The resulting `GeoDataFrame` (called *gdf*) is a collection of points, each representing a pixel from the original raster, now equipped with precise geographic locations and associated values.

```python
rows, cols = clipped_array[0].shape
col_indices, row_indices = np.meshgrid(np.arange(cols), np.arange(rows))

# Calculate the real-world coordinates
xs, ys = rasterio.transform.xy(clipped_transform, row_indices.flatten(),
                               col_indices.flatten())

# Create a DataFrame
df = pd.DataFrame({
    'X': xs,
    'Y': ys,
    'Value': clipped_array[0].flatten()
})

# Convert to a GeoDataFrame
gdf = gpd.GeoDataFrame(df, geometry=gpd.points_from_xy(df.X, df.Y))

```

```
print(gdf)
```

In the final step of this analysis, we visualize the geospatial data that has been converted into a `GeoDataFrame` of points, representing the shipping lanes in the North Sea.

A figure and axes object are created using `plt.subplots`, with the figure background set to black. The plot method of the `GeoDataFrame` (called *gdf*) is then used to render the points on the axes (ax). The column 'Value' from *gdf*, which contains the data points from the shipping lane density, is used to determine the color of each point. The size of each point is set to a very small value (*s=0.05*) to represent individual pixels effectively. The colormap 'Blues' is chosen for visual representation, and `colors.LogNorm()` is used for the normalization of color scaling, which is particularly useful for datasets with a wide range of values as it helps in highlighting variations in the data. Edges of the points are set to 'none' to ensure a cleaner visualization.

```
fig, ax = plt.subplots(facecolor='black')
fig.set_size_inches(3.5, 7)

gdf.plot(ax=ax, column='Value', s=0.05, cmap='Blues', norm=colors.LogNorm(),
         edgecolors='none')
ax.axis('off')
plt.show()
```

Figure 14.3: Shipping lanes of Northern Europe as Points

15

Mapping Population Density

15.1 Vectors to Rasters - Population Density

In this chapter, we embark on a exploration of geospatial data manipulation and visualization, focusing on the intricate relationship between vector and raster data formats. Our journey begins with the Kontour population dataset, a collection of hexagonal polygons each containing population values. Using this dataset, we will craft a detailed population density map of Colombia, showcasing not only the distribution of its inhabitants but also the versatility of geospatial data handling.

At the outset of our chapter on advanced vector and raster manipulation using population data, our first step involves preparing the reference polygon for Colombia. This reference polygon serves a dual purpose: it not only provides the necessary geographical context for our visualizations but is also pivotal in isolating the specific population data relevant to Colombia.

The process begins with loading the country boundary shapefile from the NaturalEarth dataset. With the global country boundaries loaded, we then focus on extracting the boundary specific to Colombia. This is done using a simple yet effective filtering operation:

```
colombia = df.loc[df['ADMIN'] == 'Colombia']
```

This line of code filters the `GeoDataFrame` (called *df*) to select only the row where the 'ADMIN' column matches 'Colombia'. The resulting `GeoDataFrame` called *colombia* now exclusively represents the geographical boundary of Colombia, providing us with an accurate and focused geographic area for our subsequent analyses and visualizations of population density data.

```
import geopandas as gpd
df = gpd.read_file('NaturalEarth/10m_cultural/ne_10m_admin_0_countries.shp')
colombia = df.loc[df['ADMIN'] == 'Colombia']
```

We first delve into loading and preparing the Kontour dataset for Colombia. The process begins with loading the South American portion of the Kontour dataset, which is stored in a compressed Parquet file format, known for its efficiency in handling large datasets. The code: `samerica = gpd.read_parquet("samerica.parquet.gz")` employs GeoPandas to read the Kontour dataset specific to South America. This is a straightforward operation, leveraging the *read_parquet* method, which is designed to efficiently handle Parquet files.

Reprojecting to a Compatible Coordinate Reference System (CRS): The next step involves reprojecting the loaded data to a new CRS, specifically "EPSG:4326". The code `samerica = samerica.to_crs("EPSG:4326")` achieves this, ensuring that the spatial data is in a format compatible with other datasets we plan to use, particularly the raster data we will work with later. This reprojection is critical for accurate spatial analysis and visualization, as it aligns all data layers to a common geographic framework. Finally, to extract the population data specific to Colombia, we use the `sjoin` method from GeoPandas: `colombia_kontor = gpd.sjoin(samerica, colombia, predicate="within")`.

This spatial join operation uses the within predicate to select only those hexagons from the Kontour dataset that are located within the boundaries of Colombia, as defined in the previously loaded *colombia* `GeoDataFrame`. The resulting *colombia_kontor* `GeoDataFrame` contains the Kontour population data exclusively for Colombia.

```
samerica = gpd.read_parquet("kontour.parquet.gz")
samerica = samerica[['population', 'geometry']]
samerica = samerica.to_crs("EPSG:4326")
colombia_kontor = gpd.sjoin(samerica, colombia, predicate="within")
colombia_kontor = colombia_kontor[['population', 'geometry']]
```

Next, we focus on creating a customized colormap and normalization for our population density map of Colombia using the Kontour data. We begin by establishing a range of population density values (*bounds*) that will correspond to different colors in our map. These boundaries are chosen to represent different density levels, allowing us to depict varying population concentrations across Colombia. A base colormap is derived from Matplotlib's 'inferno_r' colormap, chosen for its vibrant and distinct color gradations. The `get_cmap` function, along with `np.linspace`, is used to sample colors evenly along the 'inferno_r' gradient, corresponding to our defined boundaries. We then prepend a transparent color (*bg_color*) to these sampled colors to ensure areas with no population (or population below our minimum threshold) are rendered as transparent in the map. These colors are combined using `np.vstack` and transformed into a `ListedColormap`, *kontour_cmap,* to be used in our visualizations. The `BoundaryNorm` function is utilized to create a *kontour_norm* object. This normalization maps the population density values to the corresponding colors in our custom colormap based on the defined bounds. It ensures that each population density level is accurately represented by the appropriate color in the colormap.

```
import matplotlib.colors as colors
import numpy as np
from matplotlib import cm
from matplotlib.colors import ListedColormap

bg_color = np.array([0.0, 0.0, 0.0, 0.0])
inferno = cm.get_cmap('inferno_r', len(bounds)-1)
newcolors = inferno(np.linspace(0, 1, len(bounds)-1))
newcolors = np.vstack((bg_color, newcolors))
kontour_cmap = ListedColormap(newcolors)
kontour_norm = colors.BoundaryNorm(
    [0.0, 0.1, 1, 2, 5, 10, 20, 50, 100, 200, 1000, 2000, 10000], kontour_cmap.N)
```

We can then plot the data. Two key pieces of info here:

- `colombia_kontor.plot(...)` plots the Kontour population data onto the axes ax. The *column='population'* argument specifies that the 'population' column from the *colombia_kontor* `GeoDataFrame` is used to determine the color of each hexagon based on our previously defined colormap (*cmap=kontour_cmap*) and normalization (*norm=kontour_norm*). The *edgecolor="none"* argument removes the borders from the hexagons for a seamless visual representation.
- `colombia.boundary.plot(ax=ax, color="black", lw=0.5)` overlays the boundary of Colombia, enhancing the geographical context and providing a clear demarcation of the country's borders on the map.

```
import matplotlib.pyplot as plt

fig, ax = plt.subplots(facecolor='white')
ax.patch.set_facecolor('white')
fig.set_size_inches(7, 3.5)
colombia_kontor.plot(ax=ax, column='population',
                     cmap=kontour_cmap, norm=kontour_norm,
                     edgecolor="none")
colombia.boundary.plot(ax=ax, color="black", lw=0.5)
ax.axis('off')
plt.show()
```

The data is incredibly high resolution and this presents a problem on

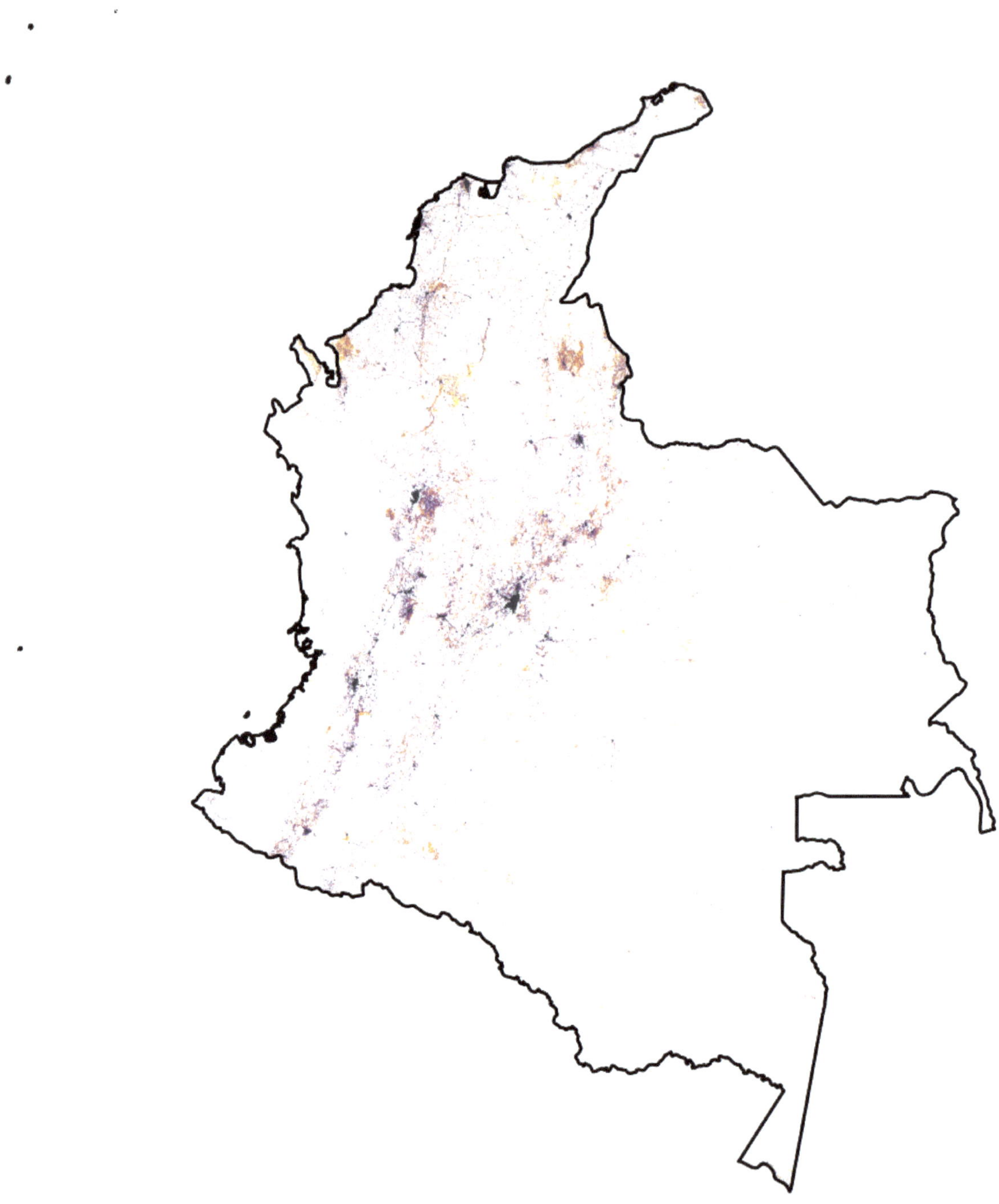

Figure 15.1: Population density of Colombia from Kontour.

a national scale. The issue of high resolution in your plot leading to a lack of visible detail is a common challenge in geospatial visualizations, especially when dealing with extensive datasets like the Kontour population data. To address this and illustrate the depth of information contained in the data, we're going to take a focused approach by zooming into a specific area. In this case, we're creating a detailed view of Medellin, a major city in Colombia, to highlight the granularity of the population data at a local level.

We start by defining the geographical coordinates of Medellin and using these to create a `Point` geometry. This point is then transformed into a `GeoDataFrame`, medellin, with the appropriate CRS ("EPSG:4326"). To capture a broader area around Medellin, we apply a buffer of 0.2 degrees to the point. This results in a circular area around the city, providing a more inclusive representation of Medellin's vicinity. The next step involves merging this localized Medellin data with the Kontour population data. Using the `sjoin` function with the predicate "within", we create a new `GeoDataFrame`, *medellin_kontor*. This spatial join operation filters the *colombia_kontor* data to include only those population hexagons that fall within the buffered area around Medellin. The result is a subset of the Kontour data, *medellin_kontor*, which specifically represents the population density around Medellin.

```
import geopandas as gpd
from shapely.geometry import Point

medellin  = gpd.GeoDataFrame({'geometry': [Point(-75.590553, 6.230833)]},
                             crs="EPSG:4326")
medellin['geometry'] = medellin['geometry'].buffer(0.2)
```

```
medellin_kontor = gpd.sjoin(colombia_kontor, medellin, predicate="within")

fig, ax = plt.subplots(facecolor='white')
ax.patch.set_facecolor('white')
fig.set_size_inches(7, 3.5)
medellin_kontor.plot(ax=ax, column='population',
                     cmap=kontour_cmap, norm=kontour_norm,
                     edgecolor="none")
ax.axis('off')
plt.show()
```

Reducing the level of detail in the map by mapping the population polygons onto a lower-resolution raster image is a strategic approach to enhance the visual comprehensibility of the data. In this process, we're using a pre-existing forests dataset as the template for the raster image, which provides a suitable resolution for your needs. Here's how this step is being executed:

Using rasterio, we load the forests dataset (forests.tif). The next step in-

Figure 15.2: Population density of Medellin from Kontour.

volves isolating the raster data to only cover the area of Colombia. This is achieved through the `rasterio.mask.mask` function, which applies a mask to the raster data based on specified geometries. The geometries used for masking are derived from the colombia `GeoDataFrame`. This has already been explained in detail in a previous chapter.

```
import rasterio
```

```
from rasterio import mask as msk
from shapely.geometry import mapping

img_forests = rasterio.open('forests.tif')

clipped_array, clipped_transform = rasterio.mask.mask(img_forests, [
    mapping(geom) for geom in (colombia.geometry.tolist())
], crop=True)

forests_array, forests_transform = rasterio.mask.mask(img_forests, [
    mapping(geom) for geom in (colombia.geometry.tolist())
], crop=True, nodata=np.amin(clipped_array))
```

In this step, we are transitioning from vector to raster representation for the population data, using the forests raster of Colombia as a template. This process, known as rasterization, involves converting the vector polygons (hexagons from the Kontour population dataset) into a raster format. Here's a detailed explanation of how this is accomplished:

Preparing Geometry and Value Pairs

The line `geom_value = ((geom, value)` for geom, value in `zip(colombia_kontor.geometry, colombia_kontor['population']))` creates a generator of tuples. Each tuple contains a geometry from the *colombia_kontor* `GeoDataFrame` (representing a hexagon) and its corresponding population value. This pairing is essential for the rasterization process, as it links each polygon (hexagon) with a specific population value, ensuring that the resulting raster correctly represents the population distribution.

Rasterizing the Vector Data

The function `features.rasterize` from the rasterio library is used to convert the vector data into a raster format:

- `geom_value` provides the function with the geometries and their associated values.
- `out_shape = forests_array[0].shape` defines the shape (dimensions) of the output raster. We use the shape of the first layer of the forests_array raster, ensuring that our population raster matches the dimensions and resolution of the forests raster.
- `transform = forests_transform` specifies the coordinate transformation of the output raster, which we align with the transformation of the forests_array to maintain geographic consistency.
- `all_touched = True` ensures that any pixel touched by a geometry is considered for rasterization, not just the pixels whose cen-

ter is within the geometry. This leads to a more inclusive and accurate representation of the population polygons.

- `fill = 0` sets the background value of the raster. Pixels not covered by any polygon will have this value, indicating no population.
- `merge_alg = MergeAlg.replace` determines how overlapping geometries are handled. 'Replace' means that the values for overlapping areas will be replaced by the value of the last geometry processed, suitable for our use case where each polygon has a distinct, non-overlapping value.

The result of this process, rasterized, is a new raster dataset where each pixel's value represents the population density based on the original Kontour vector data, but now in a format that aligns with the resolution and extent of the Colombia forests raster. This transformation is key for integrating detailed vector data into a raster format that is more manageable for certain types of analysis and visualization, particularly when dealing with large and complex datasets.

```
from rasterio import features
from rasterio.enums import MergeAlg

geom_value = ((geom,value) for geom, value in zip(colombia_kontor.geometry,
                                                  colombia_kontor['population']))

# Rasterize vector using the shape and coordinate system of the raster
rasterized = features.rasterize(geom_value,
                                out_shape = forests_array[0].shape,
                                transform = forests_transform,
                                all_touched = True,
                                fill = 0,   # background value
                                merge_alg = MergeAlg.replace)
```

Plotting the newly rasterized population data significantly enhances our understanding of Colombia's demographic landscape at a national level. The transition from high-resolution vector data to a lower-resolution raster format results in a map that is not only easier to interpret but also visually appealing.

Utilizing Rasterio's show function, the rasterized population data is displayed on the axes. The command `show(rasterized, ax=ax, transform = forests_transform, cmap = kontour_cmap, norm = kontour_norm)` accomplishes several key tasks: Rendering the rasterized data offers a more consolidated and accessible view of population densities across Colombia. The geographical alignment is maintained with `transform=forests_transform`, ensuring spatial accuracy. The colormap (*cmap=kontour_cmap*) and normalization (*norm=kontour_norm*) vividly translate population densities into a spectrum of colors, making the map not only informative but also aesthetically pleasing. Adding Colombia's boundary with `colombia.boundary.plot(ax=ax, color="black", lw=0.5)` enhances the map's clarity and context. This overlay helps anchor the raster data within the familiar contours of Colombia's borders.

The end result is a map that elegantly conveys Colombia's population distribution at a glance. The lower resolution of the rasterized data alleviates the issue of over-detailing that plagued the original high-resolution vector map, making the overall national trends in population density much more discernible. The careful blending of data transformation techniques and visualization strategies in this map not only improves its visual quality but also enhances its utility as a tool for understanding demographic patterns on a broader scale. This approach exemplifies the power of geospatial analysis in revealing new insights and perspectives, crucial in the realm of advanced geospatial analysis.

```
from rasterio.plot import show

fig, ax = plt.subplots()
show(rasterized, ax=ax, transform=forests_transform, cmap=kontour_cmap,
     norm=kontour_norm)
colombia.boundary.plot(ax=ax, color="black", lw=0.5)

ax.axis('off')
plt.show()
```

Plotting the population and forest rasters side by side is an excellent way to visually compare two different aspects of Colombia's geography: its population distribution and its forest coverage. Using a dual-pane layout, this approach allows for a direct visual comparison between the two datasets. Here's how this is achieved:

The code `fig, (ax1, ax2) = plt.subplots(nrows=1, ncols=2, figsize= (10, 4))` creates a figure with two side-by-side subplots (*axes*), ax1 and ax2, with a figure size of 10 by 4 inches. This arrangement provides a balanced view, allowing for easy comparison between the two maps. On the first axis (*ax1*), the population raster (*rasterized*) is displayed using Rasterio's show function. The custom population colormap (*kontour_cmap*) and normalization (*kontour_norm*) are applied, ensuring the population data is represented with appropriate colors and intensities. The boundary of Colombia is overlaid on the same axis using `colombia.boundary.plot(ax=ax1, color="black", lw=0.5)`, providing geographic context to the population data. The second axis (*ax2*)

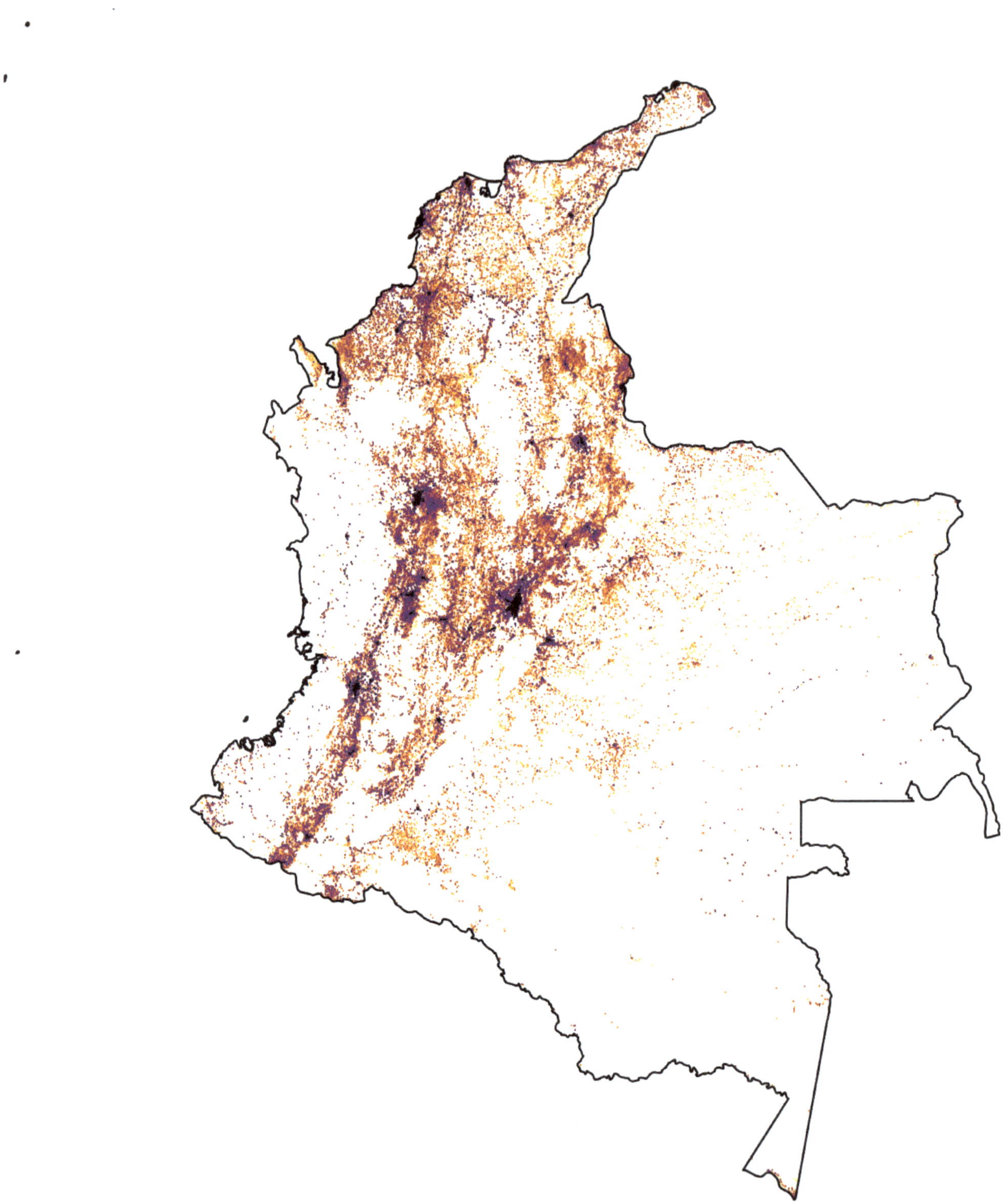

Figure 15.3: Population density of Colombia from after downsampling.

is dedicated to the forest raster (*forests_array[0]*). The same show function is used, this time applying the custom forest colormap (*forests_cmap*) and normalization (*forests_norm*) to visually represent forest coverage.As with the population map, Colombia's boundary is plotted on the forest map for consistency and context. By presenting the population and forest rasters adjacent to each other, this visualization technique effectively showcases the spatial relationship between human habitation and forest areas in Colombia. Such a side-by-side comparison not only highlights the contrasting nature of the two datasets but also provides insights into how population centers relate to forested regions, enriching our understanding of Colombia's geographical and ecological dynamics.

```
# Plot raster
fig, (ax1, ax2) = plt.subplots(nrows=2, ncols=1, figsize=(5, 8))

show(rasterized, ax=ax1, transform=forests_transform, cmap=kontour_cmap,
     norm=kontour_norm)
colombia.boundary.plot(ax=ax1, color="black", lw=0.5)
ax1.axis('off')

show(forests_array[0], ax=ax2, transform=forests_transform, cmap=forests_cmap,
     norm=forests_norm)
colombia.boundary.plot(ax=ax2, color="black", lw=0.5)
ax2.axis('off')

plt.tight_layout()
plt.show()
```

Raster to Polygons

The next step is to go the other way and convert rasters into polygons. The idea of converting a forests raster into vector polygons is a fascinating exercise in geospatial data manipulation. It demonstrates the versatility and reversibility of spatial data formats, providing a full-circle understanding of raster and vector data transformations. Here's an explanation of how this process is typically achieved in Python using relevant libraries:

Setting Up the GDAL Transform: *gdal_transform* is constructed using values from *forests_transform*, which represents the affine transformation of the raster. This transformation includes information about the pixel size and the origin coordinates, crucial for geospatial positioning of the raster.

Creating an In-Memory Raster:

- `mem_drv_raster = gdal.GetDriverByName('MEM')` initializes an in-memory raster driver. 'MEM' indicates that the raster will be

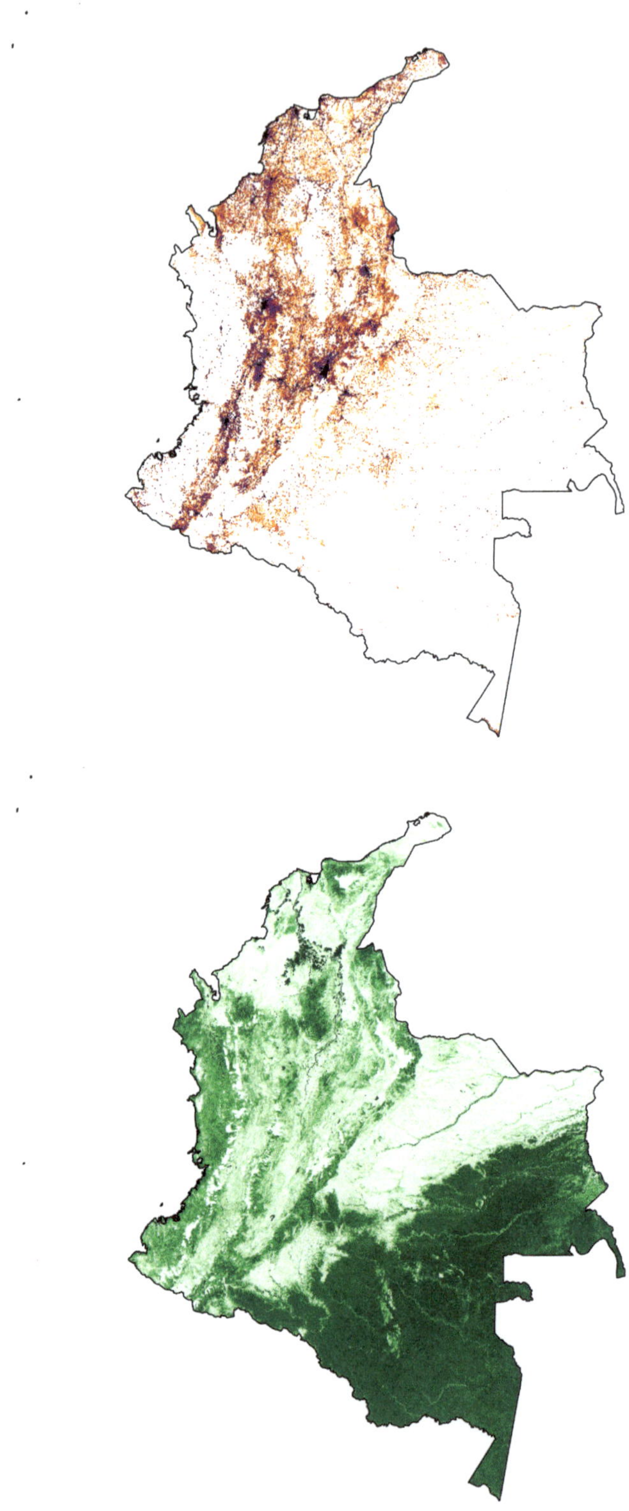

Figure 15.4: Side by side comparison of Colombian population density and forest cover.

created in memory rather than written to a file.

- `mem_raster` is then created with dimensions and data type matching the original forests_array raster (specifically, its first band). This in-memory raster serves as a temporary data store for the conversion process.

The raster object (*mem_raster*) is assigned a geotransform using the `SetGeoTransform(gdal_transform)` method. This step ensures the raster has the correct spatial reference and dimensions. The raster band is then populated with data from *forests_array[0]* using `WriteArray`. This effectively copies the forest raster data into the in-memory raster.

An in-memory vector data source (*mem_vector*) is created using the 'Memory' driver from OGR. A layer (*mem_layer*) for polygonized data is then created within this data source, with the geometry type set to polygons (*ogr.wkbPolygon*).

A new field (*new_field*) is created and added to *mem_layer*. This field is designed to store the values from the raster data (denoted as 'DN' for "Digital Number" or raster value), allowing each resulting polygon to have an attribute reflecting its original raster value.

The `gdal.Polygonize` function is used to convert the raster data into vector polygons. This function takes the raster band (`mem_raster.GetRasterBand(1)`), and polygonizes it into the *mem_layer*. Each polygon in *mem_layer* represents an area of contiguous, similar values in the raster (e.g., a contiguous forested area).

```
import numpy as np
from osgeo import gdal, ogr
from osgeo import osr
from rasterio.transform import from_origin

gdal_transform = (forests_transform.c, forests_transform.a, forests_transform.b,
                  forests_transform.f, forests_transform.d, forests_transform.e)

# Create an in-memory raster
mem_drv_raster = gdal.GetDriverByName('MEM')
mem_raster = mem_drv_raster.Create('', forests_array[0].shape[1],
                                   forests_array[0].shape[0], 1, gdal.GDT_Int32)

# Set the geotransform
mem_raster.SetGeoTransform(gdal_transform)

# Write the data
mem_raster.GetRasterBand(1).WriteArray(forests_array[0])

# Create an in-memory vector layer
mem_drv_vector = ogr.GetDriverByName('Memory')
mem_vector = mem_drv_vector.CreateDataSource('out')
mem_layer = mem_vector.CreateLayer('polygonized_raster', geom_type=ogr.wkbPolygon)
```

```

# Add a new field to store raster values
new_field = ogr.FieldDefn('DN', ogr.OFTInteger)
mem_layer.CreateField(new_field)

# Polygonize the raster to the in-memory vector layer
gdal.Polygonize(mem_raster.GetRasterBand(1), None, mem_layer, 0, [], callback=None)
```

The next section of the code extracts vector polygon geometries and their corresponding values from the in-memory layer created in the previous steps, where the forests raster was polygonized. The aim is to transform these geometries into a format that can be used in Python for further analysis or visualization. Here's a breakdown of the process:

geoms and *vals* are initialized as empty lists. These lists will store the polygon geometries (*geoms*) and their associated raster values (*vals*) respectively. The for loop (`for feature in mem_layer:`) iterates over each feature (polygon) in the *mem_layer*. Each feature represents a polygonized area of the original raster. `geom = feature.GetGeometryRef()` retrieves the geometry reference of the current feature (polygon). The code then accesses the exterior ring of the polygon with:
`ring = geom.GetGeometryRef(0)`. This is necessary because polygon geometries in OGR are composed of one exterior ring (the boundary) and potentially multiple interior rings (holes). A nested for loop collects the coordinates of each point in the ring. `ring.GetPointCount()` gives the number of points in the ring. For each point, its X and Y coordinates are retrieved using *GetX(i)* and *GetY(i)*. These coordinates are appended as tuples ((x, y)) to the coordinates list, which represents the vertices of the polygon. A Polygon object is created from the list of coordinates using `Polygon(coordinates)` and appended to the geoms list. This converts the OGR geometry into a Shapely geometry, which is more versatile and easier to work with in Python. The corresponding raster value (Digital Number - DN) for the polygon is retrieved using `feature.GetField('DN')` and appended to the vals list. *geoms*, containing Shapely polygon objects for each area in the original raster, and *vals*, holding the associated raster values for these polygons. These lists provide a convenient and Python-friendly way to work with the data previously locked in a raster format, now transformed into individual vector geometries. This conversion is a crucial step for analyses that require vector data, enabling a wide range of spatial operations and visualizations.

```
from shapely.geometry import Polygon

geoms = []
vals = []
```

```

# Access the polygons
for feature in mem_layer:
    geom = feature.GetGeometryRef()

    # Extract the coordinates from the geometry
    ring = geom.GetGeometryRef(0)
    coordinates = []
    for i in range(ring.GetPointCount()):
        x, y = ring.GetX(i), ring.GetY(i)
        coordinates.append((x, y))

    geoms.append(Polygon(coordinates))
    vals.append(feature.GetField('DN'))
```

We can then convert our geometries and values into a GeoDataFrame.

```
import pandas as pd

gdf = gpd.GeoDataFrame(pd.DataFrame({'vals': vals}), crs='EPSG:4326', geometry=geoms)
```

It is important to remember that the polygons are not the same size. We have essentially mapped areas of continuous, identical values to a single polygon.

```
gdf['area'] = gdf.area
print(gdf.sort_values('area'))
```

```
         vals                                           geometry        area
0         254  POLYGON ((-80.09167 13.57917, -80.09167 13.575...    0.000017
3934911    90  POLYGON ((-76.02917 0.46667, -76.02917 0.46250...    0.000017
3061804    84  POLYGON ((-69.45417 2.30000, -69.45417 2.29583...    0.000017
3061803   100  POLYGON ((-69.45833 2.30000, -69.45833 2.29583...    0.000017
3934910    94  POLYGON ((-76.03333 0.46667, -76.03333 0.46250...    0.000017
...       ...                                                ...         ...
1965571     0  POLYGON ((-76.52917 4.90833, -76.52083 4.90833...    0.053628
3173534     0  POLYGON ((-77.25000 2.61667, -77.24583 2.61667...    0.060469
1157874     0  POLYGON ((-72.13333 6.65417, -72.12500 6.65417...    0.065486
4635181     0  POLYGON ((-67.44583 2.13750, -67.41667 2.13750...   17.019653
4635164     0  POLYGON ((-81.72500 13.57917, -81.72500 12.512...  155.093611

[4635182 rows x 3 columns]
```

Finally, we can plot the data. It looks more or less identical to the original raster plot, however we can now do things like vector manipulations or combine with other vector datasets.

```
import matplotlib.pyplot as plt

fig, ax = plt.subplots()
fig.set_size_inches(7, 3.5)
gdf.plot(ax=ax, column='vals', cmap=forests_cmap, norm=forests_norm)
colombia.boundary.plot(ax=ax, color="black", lw=0.5)
ax.axis('off')
plt.show()
```

16

Mapping Contours

16.1 Rasters to Vectors - Contour Maps

As we embark on the chapter dedicated to transforming a Digital Elevation Model (DEM) into a contour line map of Greece, it's important to highlight a shift in our approach concerning the tools we will be utilizing. In previous chapters, our focus was on employing Rasterio, a Python library that simplifies the reading, writing, and processing of geospatial raster data. Rasterio itself is built on top of the Geospatial Data Abstraction Library (GDAL), a powerful suite of tools for processing raster and vector geospatial data.

For this chapter, however, we will be working directly with GDAL. This decision is intentional and serves a dual purpose. Firstly, it allows us to delve deeper into the capabilities and functionalities offered by GDAL, providing a more foundational understanding of the workings of geospatial data manipulation at a lower level. Secondly, by using GDAL directly, we gain more granular control over the data processing steps, which is particularly beneficial when converting DEMs to contour maps—a process that requires precision and customization.

GDAL offers robust tools specifically designed for handling DEMs and generating contour lines, making it an ideal choice for this chapter. We will explore how to use GDAL's functions to directly read the DEM of Greece, extract elevation information, and then accurately render this information as contour lines on a map. This approach not only enriches our technical skillset but also enhances our understanding of the terrain of Greece through the detailed and precise representation that contour maps provide.

By focusing on GDAL for this particular task, we aim to provide a comprehensive learning experience that balances both the theoretical aspects of DEMs and contour mapping with practical, hands-on application using one of the most powerful tools in the field of geospatial data processing. As usual we are going to load a shapefile with polygons of country boundaries and this time we will index the ones for Greece.

```
import pandas as pd
import geopandas as gpd


df = gpd.read_file('NaturalEarth/10m_cultural/ne_10m_admin_0_countries.shp')

greece = df.loc[df['ADMIN'] == 'Greece']
```

The objective here is to define the geographic area (in this case, Greece) that will serve as the boundary for our contour map generation. This involves using Python's built-in capabilities, along with GDAL, to create and handle a GeoJSON representation of Greece's boundary. Here's how this is done:

We start by defining a GeoJSON structure in Python. GeoJSON is a popular format for encoding various geographic data structures. The *geojson_data* dictionary is crafted to represent a GeoJSON `FeatureCollection`. Within this collection, we create a `Feature` that describes the geometry of Greece. This geometry is extracted from the greece `GeoDataFrame`, specifically from its first row (*iloc[0]*). The *__geo_interface__* attribute of a Shapely geometry object is a powerful tool that conveniently converts the geometry into a GeoJSON format. This method of extraction and conversion is particularly useful as it allows us to directly use the geometry data from our `GeoDataFrame` in a format that GDAL can understand.

The GeoJSON data is then serialized into a JSON string using: `json.dumps(geojson_data)`. This conversion is necessary because the next step in our process involves working with this data as a string.

GDAL's functionality is extended beyond traditional file operations, as it can also work with virtual file systems. This feature is particularly useful for operations that require temporary files. The `gdal.FileFromMemBuffer` function is used to create a virtual file in memory *("/vsimem/clipping_geom.geojson")*. This virtual file path is recognized by GDAL and allows us to work with files directly in memory without needing physical file storage. The GeoJSON string (*geojson_str*) is written to this virtual file. This step is crucial as it prepares our geographic boundary data in a format that can be used by GDAL for subsequent operations, such as clipping the DEM to the area of Greece.

By performing these steps, we effectively prepare the geographic boundary of Greece in a suitable format for processing with GDAL. This preparation is key to ensuring that our subsequent steps in generating a contour map are geographically precise and confined to our area of interest. It demonstrates the seamless integration of Python's data manipulation capabilities with GDAL's powerful geospatial processing tools, setting the stage for the intricate task of DEM analysis and contour map creation.

```
import tempfile
import json
from osgeo import gdal

geojson_data = {
    "type": "FeatureCollection",
    "features": [{
        "type": "Feature",
        "properties": {},
        "geometry": greece.geometry.iloc[0].__geo_interface__
    }]
}
geojson_str = json.dumps(geojson_data)

# Write GeoJSON string to the virtual file
gdal.FileFromMemBuffer("/vsimem/clipping_geom.geojson", geojson_str)
```

We need to merge two rasters which neighbour each other. In this portion of our chapter focused on transforming a Digital Elevation Model (DEM) into a contour map of Greece, we are proceeding with the process of creating a mosaic of the topographic data and then clipping it to the geographical boundary of Greece. The steps involved here are crucial in preparing our DEM data for contour line generation. It is important to highlight here that in our previous chapter on topography maps we used rasterio to merge rasters. Here we are going to use GDAL. The steps outlined in the previous chapter can be used here and with a few extra steps, achieve the same end result.

The first task involves creating a mosaic of topographic raster files. This is accomplished using GDAL's `BuildVRT` function. The variable file_list contains the file paths of the individual DEM raster files that cover different sections of the geographical area of interest. The `BuildVRT` function takes these individual rasters and virtually stitches them together into a single composite raster, referred to as a Virtual Raster Tile (VRT). This VRT is stored in memory at the virtual path '/vsimem/temp.vrt'. Creating a VRT is advantageous as it allows for the seamless combination of multiple rasters without the need to create a large, physical file on disk. It's an efficient way to manage and process multiple raster datasets as a unified entity.

Following the creation of the VRT, we use GDAL's Translate function to convert the VRT into a single raster file. This step is important because it consolidates the virtual mosaic into a single, manageable raster dataset. The output of this operation is another virtual file, */vsimem/mosaic.tif*, which represents the combined raster data of the specified topographic files.

The final and critical step in preparing our DEM data involves clipping this mosaic raster to the shape of Greece. This is where the virtual GeoJSON file we created earlier (representing Greece's boundary) comes into play. The `gdal.Warp` function is used for this clipping operation. It takes the mosaic raster as input and applies a 'warping' transformation that clips the raster based on the specified cutline (our GeoJSON boundary of Greece, referenced by */vsimem/clipping_geom.geojson*). The *cropToCutline=True* parameter ensures that the *output raster, clipped_raster*, is cropped to the exact boundary of Greece. The *format='MEM'* parameter indicates that the output is an in-memory raster, and *dstNodata=11* sets the 'NoData' value for the output raster, which is useful for handling areas outside the cutline.

By executing these steps, we successfully prepare a DEM of Greece that is tailored to the exact geographical boundaries of the country. This clipped raster serves as the foundational data for generating contour lines, allowing us to visualize the topographic variations across Greece with precision and clarity. This process not only demonstrates the power of GDAL in handling complex raster operations but also sets the stage for the intricate and visually rich task of contour mapping.

```
file_list = ['30N000E_20101117_gmted_mea075.tif',
             '30N030E_20101117_gmted_mea075.tif']
```

```
vrt = gdal.BuildVRT('/vsimem/temp.vrt', file_list)
mosaic = gdal.Translate('/vsimem/mosaic.tif', vrt)

# Clip the raster using the geometry
clipped_raster = gdal.Warp('', mosaic, format='MEM',
                           cutlineDSName="/vsimem/clipping_geom.geojson",
                           cropToCutline=True, dstNodata=11)
```

In this stage of our chapter on converting a Digital Elevation Model (DEM) of Greece into a contour line map, we focus on extracting the necessary geospatial and pixel data from the clipped raster. This involves understanding the raster's geospatial transformation and accessing its pixel values. Let's break down these steps:

The geotransform of a raster in GDAL defines how pixel coordinates are mapped to geographic coordinates. It includes information about the raster's origin (top-left corner), the pixel size, and rotation. The line `gdal_transform = clipped_raster.GetGeoTransform()` extracts this geotransform from the
clipped_raster, which is our raster dataset clipped to the boundary of Greece. This transform is essential for any spatial referencing and calculations we might need to perform on the raster data.

While GDAL's transform is effective, working with it directly in Python can sometimes be cumbersome. To facilitate easier manipulation and interaction with the transform, we convert it to an Affine transform. `Affine.from_gdal(*gdal_transform)` achieves this conversion.
The Affine library provides a more Python-friendly way of dealing with transformations, allowing for straightforward operations like scaling, rotating, and translating geometries. The resulting affine_transform object represents the same spatial information as the original GDAL transform but in a format that's more conducive to Python operations.

Finally, we access the actual pixel values of the clipped raster. The raster band (in most DEMs, elevation data is stored in the first band) is accessed via `clipped_raster.GetRasterBand(1).ReadAsArray()` is then called on this band to read its data into a NumPy array, called *clipped_array*. This array now contains the elevation values for each pixel in the raster, corresponding to the geographic area of Greece.

By completing these steps, we now have both the spatial reference (through the Affine transform) and the elevation data (in the NumPy array) necessary for generating contour lines. This data will be instrumental in the next phase of our analysis, where we will use these elevation values to create and visualize the contours that represent Greece's topography. This process not only exemplifies the technical process of data extraction from rasters but also sets the foundation for the detailed topographical visualization we aim to achieve in our contour map of Greece.

```
import matplotlib.pyplot as plt
from rasterio.plot import show
from affine import Affine


# Get the GDAL transform
gdal_transform = clipped_raster.GetGeoTransform()
affine_transform = Affine.from_gdal(*gdal_transform)
```

```


# Read data from the clipped raster
clipped_array = clipped_raster.GetRasterBand(1).ReadAsArray()
# Plot the data
plt.figure(figsize=(8, 8))
show(clipped_array, transform=affine_transform, extent=[0,
     clipped_array.shape[1], 0, clipped_array.shape[0]])
plt.show()
```

```
from osgeo import ogr


# Generate contours directly in memory
mem_ds = ogr.GetDriverByName('Memory').CreateDataSource('out')
contour_lyr = mem_ds.CreateLayer('contour', geom_type=ogr.wkbLineString25D)

# Define fields of id and elev
field_def = ogr.FieldDefn("ID", ogr.OFTInteger)
contour_lyr.CreateField(field_def)
field_def = ogr.FieldDefn("elev", ogr.OFTReal)
contour_lyr.CreateField(field_def)

# Generate contours
gdal.ContourGenerate(clipped_raster.GetRasterBand(1), 10, -11, [], 0, 0,
                     contour_lyr, 0, 1)

gdal.Unlink('/vsimem/temp.vrt')
gdal.Unlink('/vsimem/mosaic.tif')
```

In this final step of our chapter, we focus on generating contour lines from the clipped raster representing Greece's topography. This process involves using GDAL and OGR libraries to create contours directly in memory, showcasing an efficient approach to handling spatial data. Let's explore these steps in detail:

The code begins by creating an in-memory data source using OGR, a part of the GDAL suite designed for vector data manipulation. The `ogr.GetDriverByName('Memory').CreateDataSource('out')` command initializes an in-memory data source (*mem_ds*), which serves as a temporary storage for the contour lines we will generate. This in-memory approach allows for faster processing and avoids the need for writing to and reading from physical files.

Within this data source, a new layer (*contour_lyr*) is created specifically for storing the contour lines. The geometry type of this layer is set to *ogr.wkbLineString25D*, which is suitable for storing line geometries with elevation (2.5D). The contours will be represented as line strings, with each line string capturing a contour line at a certain elevation.

Two fields are defined and added to the contour layer: 'ID' (an integer field) and 'elev' (a real number field). The 'ID' field will store a unique identifier for each contour line, while the 'elev' field will store

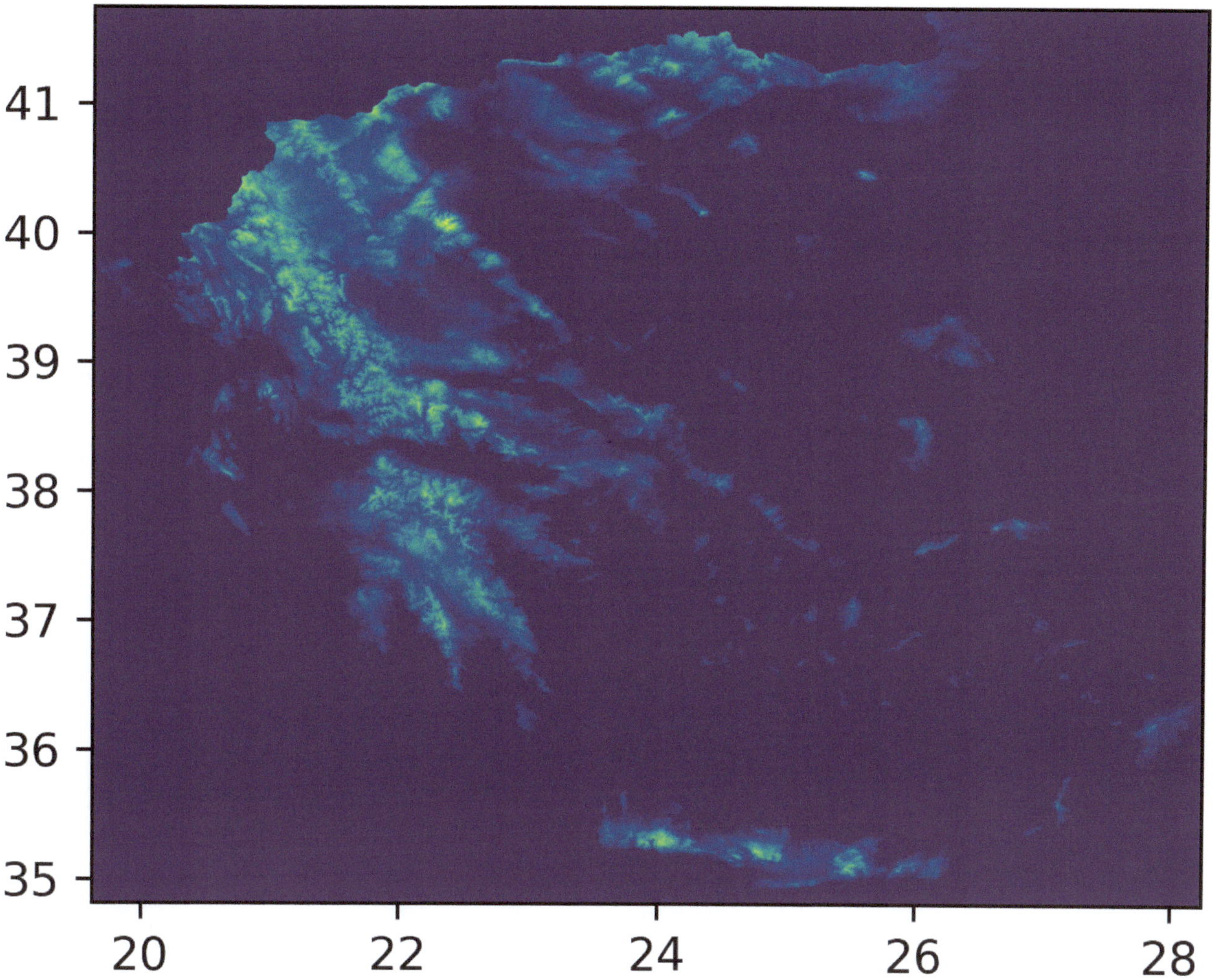

Figure 16.1: Topography of Greece

the elevation value associated with each contour line.

The `gdal.ContourGenerate` function is used to generate the contours from the clipped raster. This function takes the raster band (*clipped_raster.GetRasterBand(1)*), contour interval (10 in this case), base contour level (-11), and other parameters, and populates the *contour_lyr* layer with contour lines. The contour interval of 10 means that a new contour line is generated for every 10 units of elevation change. The base contour level is set to -11, which can be adjusted based on the elevation range of the dataset.

After generating the contours, the virtual files created earlier in the process (*/vsimem/temp.vrt* and */vsimem/mosaic.tif*) are unlinked or deleted using gdal.Unlink. This step is important for managing memory and ensuring no residual data is left in the virtual file system.

```
from shapely.geometry import LineString

lines = [LineString(np.array(contour.GetGeometryRef().GetPoints()))
         for contour in contour_lyr]
elev = [contour.GetField('elev') for contour in contour_lyr]
```

We next need to convert these contour lines into a usable geospatial format. The process starts by iterating over each contour in the *contour_lyr*, which is the layer containing the contour lines we generated earlier. For each contour line, the geometry is accessed by calling contour.GetGeometryRef().GetPoints() on the geometry reference to retrieve the coordinates of all the points that make up the contour line. These points are then used to create a LineString object from the Shapely library. Along with the geometries, we also extract the elevation data associated with each contour line. This is achieved by iterating over the same *contour_lyr* and using contour.GetField('elev') to obtain the elevation value ('elev') for each contour. These elevation values are crucial as they provide the vertical dimension to our contour lines, indicating the altitude each line represents on the ground. *lines*, containing the LineString representations of the contour lines, and elev, holding the corresponding elevation values. These lists can then be used to construct a GeoDataFrame in GeoPandas, which combines the spatial geometry of the contour lines with their associated elevation data. This GeoDataFrame will serve as a versatile and powerful tool for further spatial analysis, visualization, and exploration of Greece's topographical features, as represented by the contour lines.

```
import geopandas as gpd

gdf = gpd.GeoDataFrame(pd.DataFrame({'elev': elev}), crs="EPSG:4326", geometry=lines)

print(gdf)
```

Finally converting it to a GeoDataFrame.

```
from matplotlib import cm
from matplotlib.colors import BoundaryNorm

cmap = cm.get_cmap("inferno", int(abs(np.amin(gdf.elev))) + int(abs(np.amax(gdf.elev))))
bounds = np.arange(int(abs(np.amin(gdf.elev))), int(abs(np.amax(gdf.elev))), 1)
norm = BoundaryNorm(bounds, cmap.N)
```

```
          elev                                           geometry
0         49.0  LINESTRING Z (26.22382 41.74986 49.00000, 26.2...
1         49.0  LINESTRING Z (26.21132 41.74361 49.00000, 26.2...
2         49.0  LINESTRING Z (26.21340 41.74986 49.00000, 26.2...
3         49.0  LINESTRING Z (26.22382 41.73979 49.00000, 26.2...
4         49.0  LINESTRING Z (26.22174 41.73736 49.00000, 26.2...
...        ...                                                ...
194243   149.0  LINESTRING Z (24.10665 34.81653 149.00000, 24....
```

```
194244  159.0  LINESTRING Z (24.10676 34.81653 159.00000, 24....
194245  169.0  LINESTRING Z (24.10688 34.81653 169.00000, 24....
194246  179.0  LINESTRING Z (24.10699 34.81653 179.00000, 24....
194247  189.0  LINESTRING Z (24.10711 34.81653 189.00000, 24....

[194248 rows x 2 columns]
```

Finally we can visualize the GeoDataFrame (*gdf*) on a map with a custom color map based on elevation data ('elev' column):

`cm.get_cmap("inferno", int(abs(np.amin(gdf.elev))) + int(abs(np.amax(gdf.elev))))` creates a colormap using Matplotlib's 'inferno' color scheme. The number of colors in the colormap is determined by the range of elevation values in gdf, ensuring a distinct color for each elevation unit.

`bounds = np.arange(int(abs(np.amin(gdf.elev))), int(abs(np.amax(gdf.elev))), 1)` sets the boundaries for color normalization. The range goes from the minimum to the maximum elevation in gdf, with an interval of 1, matching the granularity of the elevation data.

`norm = BoundaryNorm(bounds, cmap.N)` creates a normalization object that maps elevation values to the corresponding indices in the colormap. It ensures that elevation values are accurately represented by colors.

```
fig, ax = plt.subplots()
gdf.plot(ax=ax, lw=0.01, column='elev', norm=norm, cmap=cmap)
ax.axis('off')
plt.show()
```

`gdf.plot(ax=ax, lw=0.01, column='elev', norm=norm, cmap=cmap)` plots the `GeoDataFrame` using the specified linewidth (*lw*), elevation column for coloring (column='elev'), and the defined colormap and normalization. This results in a map where different elevations are represented by different colors, providing a clear visual representation of the terrain's elevation profile.

In the previous section, we transformed a raster dataset into a GeoDataFrame of contour lines, converting continuous raster data into discrete vector representations. This process allowed us to extract meaningful elevation contours, which are valuable for various analyses and visualizations.

Now, we'll reverse this process by converting our GeoDataFrame of contour lines back into a raster format. This step is crucial when we need to

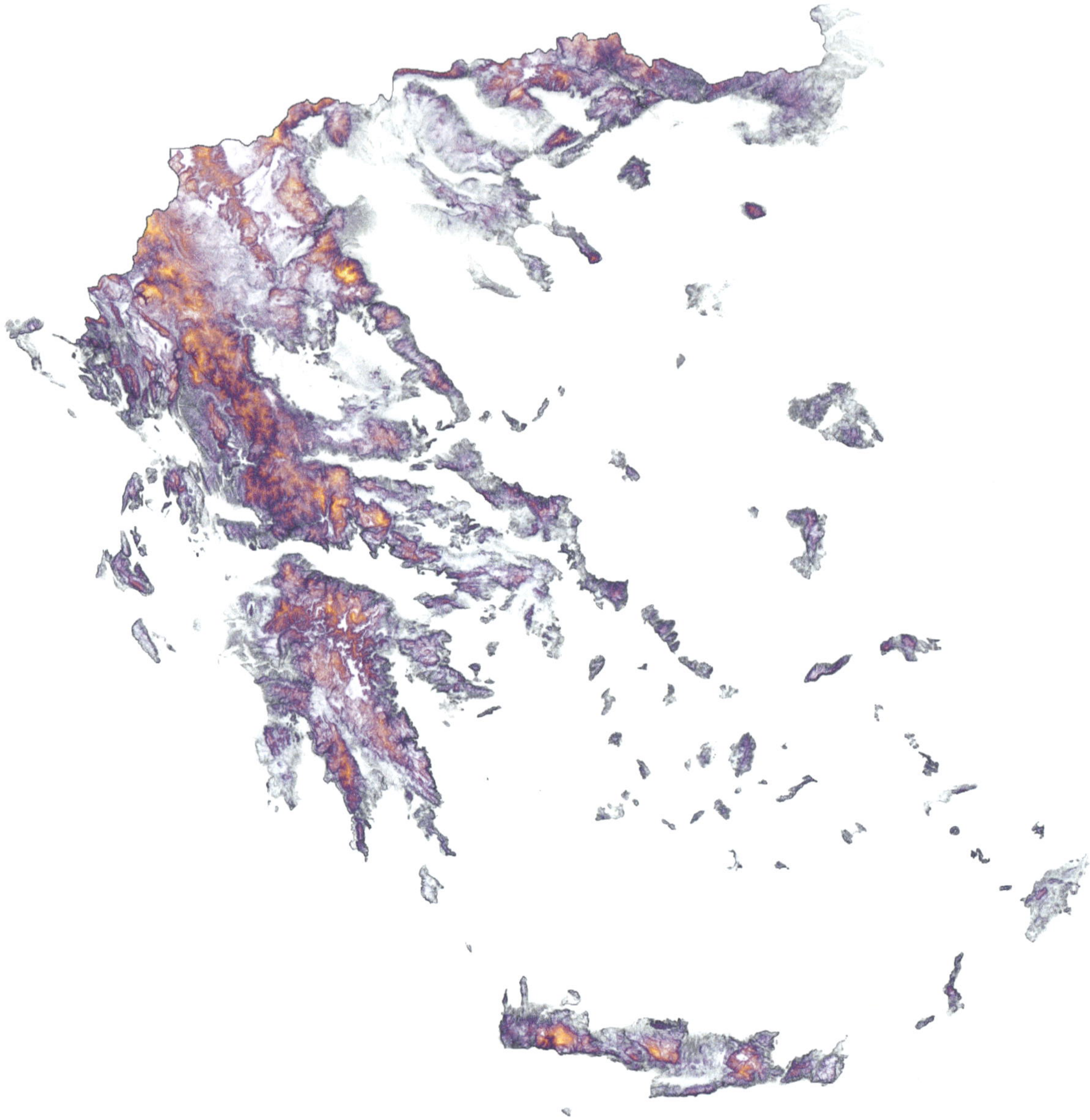

Figure 16.2: Contour line map of Greece.

perform raster-based analyses or when raster visualization better suits our data representation needs.

To begin, we need to define the spatial extent and resolution of the output raster to ensure it aligns correctly with our contour lines. We extract the minimum and maximum coordinates from our GeoDataFrame using minx, miny, maxx, maxy = gdf.total_bounds. This function returns the smallest and largest x and y coordinates among all geometries in the GeoDataFrame, effectively giving us the bounding box that encompasses all the contour lines.

Next, we set a pixel_size of 0.001, which determines the resolution of the output raster. A smaller pixel size results in a higher-resolution raster with more detail but increases computational load and file size. Conversely, a larger pixel size creates a lower-resolution raster, which is less detailed but more efficient to process. Choosing an appropriate pixel size is essential; it should be small enough to capture the necessary detail from the contour lines but also consider processing efficiency.

We then calculate the raster's dimensions with width = int((maxx - minx) / pixel_size) and height = int((maxy - miny) / pixel_size). This determines how many pixels will fit along each dimension of the raster, given the desired resolution. We convert these values to integers because raster dimensions must be whole numbers. These dimensions define the size of the raster grid that will cover the entire spatial extent of our contour lines at the specified resolution.

To avoid any potential errors, we ensure that the width and height are at least one pixel in size by using width = max(width, 1) and height = max(height, 1). This is a safeguard against cases where the spatial extent is smaller than the pixel size, which could result in dimensions of zero pixels. Without this check, the rasterization process would fail because a raster cannot have zero dimensions.

```
minx, miny, maxx, maxy = gdf.total_bounds
pixel_size = 0.001

width = int((maxx - minx) / pixel_size)
height = int((maxy - miny) / pixel_size)

width = max(width, 1)
height = max(height, 1)
```

To proceed with the rasterization process, we first need to create an affine transformation that maps pixel coordinates to spatial coordinates. This transformation is essential because it defines how the raster grid aligns with the spatial extent of our data. We use the from_bounds function from the rasterio.transform module, providing it with the minimum and maximum coordinates of our GeoDataFrame (minx, miny, maxx, maxy) and the dimensions of the raster grid (width, height)

With the affine transform established, we prepare the geometries and their associated values for rasterization. Our goal is to assign specific elevation values from our contour lines to the corresponding pixels in the raster. We achieve this by creating a generator expression that pairs each geometry with its elevation value from the 'elev' column in our GeoDataFrame:

This shapes generator is a sequence of tuples where each tuple contains a geometry and its associated elevation value. It serves as the input for the rasterization function, ensuring that each contour line is represented in the raster with the correct elevation.

Next, we perform the rasterization using the rasterize function from the rasterio.features module. This function converts our vector geometries into a rasterized format based on the parameters we provide.

Here's what each parameter means:

- shapes: The generator we created that pairs geometries with their elevation values.
- out_shape: The dimensions of the output array, defined by the raster's height and width.
- transform: The affine transformation that aligns the raster grid with the spatial data.
- fill: The value assigned to pixels that do not correspond to any geometry; in this case, we use 0 to represent the absence of data.
- dtype: The data type of the output array. We use 'float32' to accurately represent elevation values, which may include decimals.

The rasterize function processes each geometry and fills the corresponding pixels in the output array with the specified elevation values. Pixels not covered by any geometry receive the background fill value.

Finally, we have a two-dimensional NumPy array called raster that represents our rasterized data. Each element in this array corresponds to a pixel in the raster grid, holding either an elevation value from the contour lines or the background value. To inspect the resulting array, we can print it:

```
from rasterio.features import rasterize
from rasterio.transform import from_bounds

transform = from_bounds(minx, miny, maxx, maxy, width, height)

shapes = ((geom, value) for geom, value in zip(gdf.geometry, gdf['elev']))

raster = rasterize(
    shapes,
```

```
    out_shape=(height, width),
    transform=transform,
    fill=0,
    dtype='float32'
)
```

Finally. We can plot the raster and you can see that we are back to our starting point. Feel free to use some of the lessons from chapter 10 to customise this plot further.

```
fig = plt.figure(figsize=(8, 8))
fig.set_size_inches(7, 3.5)
ax = plt.axes()
show(raster, transform=transform, ax=ax)
plt.show()
```

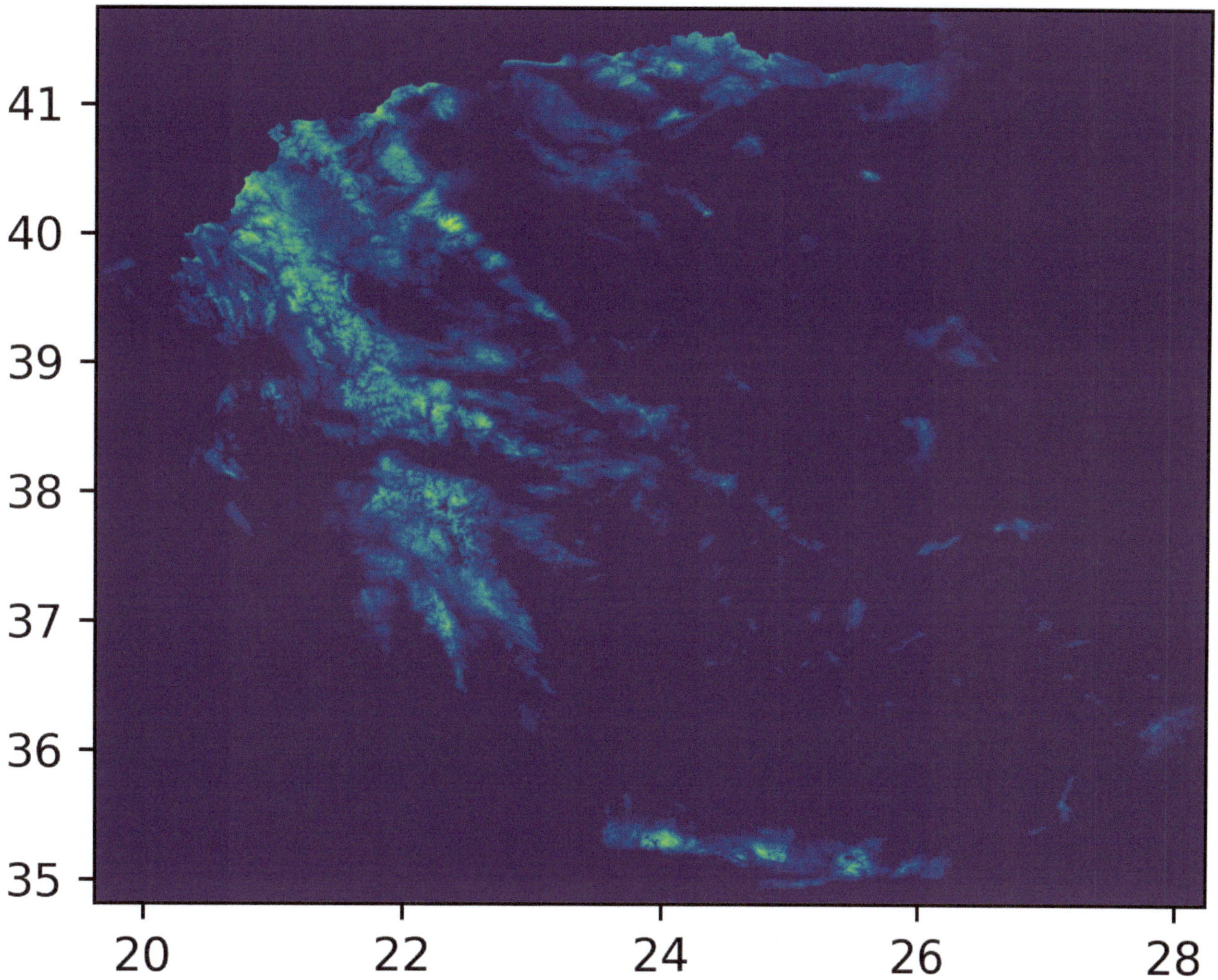

Figure 16.3: Rasterised representation of a GeoDataFrame of linestrings

Figure 16.4: A customised version.

17

Conclusion

As we reach the culmination of our journey through the intricate landscape of geospatial data visualization in Python, it's time to reflect on the ground we've covered and the tools we've acquired. This book was crafted with the intent to empower you with the knowledge and skills necessary to navigate the vast world of geospatial data—whether you're a student embarking on a new academic venture or a professional aiming to enhance your technical repertoire.

17.1 Building a Strong Foundation

We began by laying a solid foundation, introducing the fundamental concepts of geospatial data types—points, linestrings, polygons, and rasters. Through plotting ports and airports, we grasped the basics of visualizing point data. Mapping European roads allowed us to delve into linestrings, understanding how linear features are represented and rendered. The exploration of world empires provided insights into polygons, teaching us how to depict areas and boundaries on a map. By visualizing livestock populations, we ventured into raster data, learning how continuous data surfaces are managed and displayed.

17.2 Advancing Our Skills

With the basics in place, we progressed to more complex topics. The CO_{2} emissions project challenged us to customize point data visualizations, enhancing their informativeness and aesthetic appeal. Tropical storms served as a gateway to advanced plotting techniques, particularly in demonstrating various map projections and their implications on data interpretation. Our journey to Mars, visualizing Martian topography, showcased the power of manipulating color maps to highlight specific features, emphasizing how color choices can influence the readability and impact of a map.

17.3 Mastering Advanced Techniques

The latter chapters pushed the boundaries of traditional mapping. Earth's topography introduced us to the art of clipping rasters and applying hillshade effects, adding depth and realism to our visualizations. The forests project taught us the importance of reprojecting data to ensure spatial accuracy and how to clip data to focus on areas of interest. Accessing OpenStreetMap data for our lighthouses map highlighted the vast resources available and how buffering and transparency (alpha) can enhance the storytelling aspect of a map.

17.4 Data Transformation and Analysis

We didn't stop at visualization; we delved into data transformation and analysis. The shipping lanes project demonstrated how to convert point data to raster formats and back, a crucial skill for various analytical processes. Mapping population density required us to manipulate polygon data, transforming it into rasters for analysis before reverting it back for visualization. Finally, extracting contour lines from raster data equipped us with techniques to represent elevation and other continuous variables in a discrete, yet informative manner.

17.5 Embracing the Tools of the Trade

Throughout this journey, we've harnessed the power of essential Python libraries such as GDAL, Rasterio, GeoPandas, Shapely, and Cartopy. Each library served as a pillar, supporting our projects and expanding our capabilities. By integrating these tools, we've not only created visually compelling maps but also performed complex geospatial analyses, reinforcing the symbiotic relationship between data processing and visualization.

17.6 A New Perspective on Geospatial Data

By working through diverse datasets—from the topography of distant planets to the intricate networks of roads and shipping lanes—we've gained a multifaceted understanding of geospatial data. This book aimed to show that maps are more than just navigational aids; they are powerful tools for storytelling, analysis, and presenting complex data in an accessible format.

17.7 Your Journey Continues

While this book provides a comprehensive starting point, the field of geospatial data science is ever-evolving. New datasets emerge, tools are updated, and innovative visualization techniques are developed. I encourage you to continue experimenting with the skills you've acquired. Seek out new data, challenge yourself with complex projects, and perhaps most importantly, share your creations with the community.

17.8 Final Thoughts

The ability to visualize geospatial data effectively is an invaluable skill in today's data-driven world. Whether you're tracking environmental changes, planning urban developments, or exploring historical events, the maps you create can influence decisions and inspire action. Remember that every map tells a story, and with the knowledge you've gained, you're now equipped to be both a storyteller and an analyst.

Thank you for embarking on this journey. I hope it has been as enlightening for you to read as it was for me to write. The world of geospatial data is vast and filled with possibilities—may this be just the beginning of your exploration.

About Locate Press

LOCATE PRESS
OPEN SOURCE GEOSPATIAL | BOOKS

Locate Press is a book publisher, focusing on the open source geospatial niche. Many traditional publishers see geospatial books as either scientific content or geared primarily toward consumers. Unfortunately, this means they don't give them the long term care they truly deserve. With more and more technical users using open source geospatial technology (for a wide variety of reasons!), now, more than ever, you need comprehensive and reliable education and training resources.

You've come to the right place!

We know that niche is not a swear word, but a marketplace that needs serious support. Geospatial data management is a core technology for government and business, making practical teaching materials for industry and higher education crucial. We also know that reliable availability of material is key. Our books, once available, remain available long after the first few thousand are sold so that you can depend on them for course material and reference long into the future.

If you are an educator looking for high quality curriculum, we would like to hear from you. Aside from training books, we also aim to provide workshop guides and exercise booklets that you can use in your courses!

Academia is not the only place for learning and training, so Locate Press supports consultants delivering workshops and seminars. If you have solid, practical material that needs some professional polish, give us a call. Likewise, if you need bulk orders to serve your students, or to resell, we can help there too.

Locate Press was founded by Tyler Mitchell in 2012, it's flagship book being *The Geospatial Desktop* by Gary Sherman. In 2013 Gary Sherman took over the helm as publisher, guiding the company until 2021 when Locate Press returned to Tyler.

Writing for Locate Press

Are you passionate about open source software? Have an uncontrollable urge to share your knowledge with the world?

At Locate Press we're looking for books that open up the world of geospatial. We love concise, targeted titles that help people expand their knowledge and get up to speed quickly. That being said, we don't go around with blinders on—we're open to other leading edge topics related to open source.

We help put your ideas into book form, getting your expertise on paper and in print. Don't let the process scare you, we're here to guide and help all along the way—from outline to print-ready copy.

ORDER DIRECT AND SAVE UP TO 30%

Our paperbacks can be ordered directly from us with bulk discounts on 5, 10, and 25+ units: store.locatepress.com

We print in countries that are closest to you and can deliver almost anywhere. Our print books also sell through Amazon or Ingram.

E-books (PDF) are ordered and downloaded directly from locatepress.com/ebooks

Educators contact us for desk/review copies:
+1 (250) 303-1831 or
tyler@locatepress.com

www.ingramcontent.com/pod-product-compliance
Ingram Content Group UK Ltd.
Pitfield, Milton Keynes, MK11 3LW, UK
UKHW060016051026
15591UKWH00002B/233
* 9 7 8 1 9 9 8 4 1 4 0 7 9 *